Lecture Notes in Mathematics

For information about Vols. 1–729, please contact your bookseller or Springer-Verlag.

Vol. 730: Functional Differential Equations and Approximation of Fixed Points. Proceedings, 1978. Edited by H.-O. Peitgen and H.-O. Walther. XV, 503 pages. 1979.

Vol. 731: Y. Nakagami and M. Takesaki, Duality for Crossed Products of von Neumann Algebras. IX, 139 pages. 1979.

Vol. 732: Algebraic Geometry. Proceedings, 1978. Edited by K. Lønsted. IV, 658 pages. 1979.

Vol. 733: F. Bloom, Modern Differential Geometric Techniques in the Theory of Continuous Distributions of Dislocations. XII, 206 pages. 1979.

Vol. 734: Ring Theory, Waterloo, 1978. Proceedings, 1978. Edited by D. Handelman and J. Lawrence. XI, 352 pages. 1979.

Vol. 735: B. Aupetit, Propriétés Spectrales des Algèbres de Banach. XII, 192 pages. 1979.

Vol. 736: E. Behrends, M-Structure and the Banach-Stone Theorem. X, 217 pages. 1979.

Vol. 737: Volterra Equations. Proceedings 1978. Edited by S.-O. Londen and O. J. Staffans. VIII, 314 pages. 1979.

Vol. 738: P. E. Conner, Differentiable Periodic Maps. 2nd edition, IV, 181 pages. 1979.

Vol. 739: Analyse Harmonique sur les Groupes de Lie II. Proceedings, 1976–78. Edited by P. Eymard et al. VI, 646 pages. 1979.

Vol. 740: Séminaire d'Algèbre Paul Dubreil. Proceedings, 1977–78. Edited by M.-P. Malliavin. V, 456 pages. 1979.

Vol. 741: Algebraic Topology, Waterloo 1978. Proceedings. Edited by P. Hoffman and V. Snaith. XI, 655 pages. 1979.

Vol. 742: K. Clancey, Seminormal Operators. VII, 125 pages. 1979.

Vol. 743: Romanian-Finnish Seminar on Complex Analysis. Proceedings, 1976. Edited by C. Andreian Cazacu et al. XVI, 713 pages. 1979.

Vol. 744: I. Reiner and K. W. Roggenkamp, Integral Representations. VIII, 275 pages. 1979.

Vol. 745: D. K. Haley, Equational Compactness in Rings. III, 167 pages. 1979.

Vol. 746: P. Hoffman, τ-Rings and Wreath Product Representations. V, 148 pages. 1979.

Vol. 747: Complex Analysis, Joensuu 1978. Proceedings, 1978. Edited by I. Laine, O. Lehto and T. Sorvali. XV, 450 pages. 1979.

Vol. 748: Combinatorial Mathematics VI. Proceedings, 1978. Edited by A. F. Horadam and W. D. Wallis. IX, 206 pages. 1979.

Vol. 749: V. Girault and P.-A. Raviart, Finite Element Approximation of the Navier-Stokes Equations. VII, 200 pages. 1979.

Vol. 750: J. C. Jantzen, Moduln mit einem höchsten Gewicht. III, 195 Seiten. 1979.

Vol. 751: Number Theory, Carbondale 1979. Proceedings. Edited by M. B. Nathanson. V, 342 pages. 1979.

Vol. 752: M. Barr, *-Autonomous Categories. VI, 140 pages. 1979.

Vol. 753: Applications of Sheaves. Proceedings, 1977. Edited by M. Fourman, C. Mulvey and D. Scott. XIV, 779 pages. 1979.

Vol. 754: O. A. Laudal, Formal Moduli of Algebraic Structures. III, 161 pages. 1979.

Vol. 755: Global Analysis. Proceedings, 1978. Edited by M. Grmela and J. E. Marsden. VII, 377 pages. 1979.

Vol. 756: H. O. Cordes, Elliptic Pseudo-Differential Operators – An Abstract Theory. IX, 331 pages. 1979.

Vol. 757: Smoothing Techniques for Curve Estimation. Proceedings, 1979. Edited by Th. Gasser and M. Rosenblatt. V, 245 pages. 1979.

Vol. 758: C. Năstăsescu and F. Van Oystaeyen; Graded and Filtered Rings and Modules. X, 148 pages. 1979.

Vol. 759: R. L. Epstein, Degrees of Unsolvability: Structure and Theory. XIV, 216 pages. 1979.

Vol. 760: H.-O. Georgii, Canonical Gibbs Measures. VIII, 190 pages. 1979.

Vol. 761: K. Johannson, Homotopy Equivalences of 3-Manifolds with Boundaries. 2, 303 pages. 1979.

Vol. 762: D. H. Sattinger, Group Theoretic Methods in Bifurcation Theory. V, 241 pages. 1979.

Vol. 763: Algebraic Topology, Aarhus 1978. Proceedings, 1978. Edited by J. L. Dupont and H. Madsen. VI, 695 pages. 1979.

Vol. 764: B. Srinivasan, Representations of Finite Chevalley Groups. XI, 177 pages. 1979.

Vol. 765: Padé Approximation and its Applications. Proceedings, 1979. Edited by L. Wuytack. VI, 392 pages. 1979.

Vol. 766: T. tom Dieck, Transformation Groups and Representation Theory. VIII, 309 pages. 1979.

Vol. 767: M. Namba, Families of Meromorphic Functions on Compact Riemann Surfaces. XII, 284 pages. 1979.

Vol. 768: R. S. Doran and J. Wichmann, Approximate Identities and Factorization in Banach Modules. X, 305 pages. 1979.

Vol. 769: J. Flum, M. Ziegler, Topological Model Theory. X, 151 pages. 1980.

Vol. 770: Séminaire Bourbaki vol. 1978/79 Exposés 525–542. IV, 341 pages. 1980.

Vol. 771: Approximation Methods for Navier-Stokes Problems. Proceedings, 1979. Edited by R. Rautmann. XVI, 581 pages. 1980.

Vol. 772: J. P. Levine, Algebraic Structure of Knot Modules. XI, 104 pages. 1980.

Vol. 773: Numerical Analysis. Proceedings, 1979. Edited by G. A. Watson. X, 184 pages. 1980.

Vol. 774: R. Azencott, Y. Guivarc'h, R. F. Gundy, Ecole d'Eté de Probabilités de Saint-Flour VIII-1978. Edited by P. L. Hennequin. XIII, 334 pages. 1980.

Vol. 775: Geometric Methods in Mathematical Physics. Proceedings, 1979. Edited by G. Kaiser and J. E. Marsden. VII, 257 pages. 1980.

Vol. 776: B. Gross, Arithmetic on Elliptic Curves with Complex Multiplication. V, 95 pages. 1980.

Vol. 777: Séminaire sur les Singularités des Surfaces. Proceedings, 1976-1977. Edited by M. Demazure, H. Pinkham and B. Teissier. IX, 339 pages. 1980.

Vol. 778: SK1 von Schiefkörpern. Proceedings, 1976. Edited by P. Draxl and M. Kneser. II, 124 pages. 1980.

Vol. 779: Euclidean Harmonic Analysis. Proceedings, 1979. Edited by J. J. Benedetto. III, 177 pages. 1980.

Vol. 780: L. Schwartz, Semi-Martingales sur des Variétés, et Martingales Conformes sur des Variétés Analytiques Complexes. XV, 132 pages. 1980.

Vol. 781: Harmonic Analysis Iraklion 1978. Proceedings 1978. Edited by N. Petridis, S. K. Pichorides and N. Varopoulos. V, 213 pages. 1980.

Vol. 782: Bifurcation and Nonlinear Eigenvalue Problems. Proceedings, 1978. Edited by C. Bardos, J. M. Lasry and M. Schatzman. VIII, 296 pages. 1980.

Vol. 783: A. Dinghas, Wertverteilung meromorpher Funktionen in ein- und mehrfach zusammenhängenden Gebieten. Edited by R. Nevanlinna and C. Andreian Cazacu. XIII, 145 pages. 1980.

Vol. 784: Séminaire de Probabilités XIV. Proceedings, 1978/79. Edited by J. Azéma and M. Yor. VIII, 546 pages. 1980.

Vol. 785: W. M. Schmidt, Diophantine Approximation. X, 299 pages. 1980.

Vol. 786: I. J. Maddox, Infinite Matrices of Operators. V, 122 pages. 1980.

continuation on page 203

Lecture Notes in Mathematics

Edited by A. Dold and B. Eckmann

953

Iterative Solution of Nonlinear Systems of Equations

Proceedings of a Meeting held at
Oberwolfach, Germany, Jan. 31 – Feb. 5, 1982

Edited by R. Ansorge, Th. Meis, and W. Törnig

Springer-Verlag
Berlin Heidelberg GmbH 1982

Editors
Rainer Ansorge
Institut für Angewandte Mathematik, Universität Hamburg
Bundesstr. 55, 2000 Hamburg 13, Germany

Theodor Meis
Mathematisches Institut, Universität Köln
Weyertal 86-90, 5000 Köln 41, Germany

Willi Törnig
Fachbereich Mathematik, TH Darmstadt
Schloßgartenstr. 7, 6100 Darmstadt, Germany

AMS Subject Classifications (1980): 65 B 05, 65 F 10, 65 F 15, 65 G 10, 65 H 10, 65 H 15, 65 N 05, 65 N 30, 70 K 10, 73 D 30, 76-04, 76 D 05, 76 N 10, 76 S 05

ISBN 978-3-540-11602-8 ISBN 978-3-540-39379-5 (eBook)
DOI 10.1007/978-3-540-39379-5

Originally published by Springer-Verlag Berlin Heidelberg New York in 1982

2146/3140-543210

FOREWORD

The meeting on Iterative Solution of Nonlinear Systems of Equations, held in the Mathematisches Forschungsinstitut Oberwolfach, Federal Republic of Germany, during the six days of January 31st to February 5th 1982, was attented by forty one mathematicians and engineers from several countries. In all twenty four lectures were given, thirteen of which are presented in these proceedings.

Emphasis was on three main topics: multigrid methods, monotone and interval arithmetic iterations, and applications in industrial practice.

Several contributors reported on the effective use of multigrid algorithms even in bifurcation and other highly nonlinear problems. The principle of error inclusion by means of interval arithmetics and monotone iterations has been investigated for several years. Recent advances in accelerating those iterations and some connections with the question of global convergence were reported on at the meeting. Finally there were stimulating contributions and discussions on concrete numerical problems in aerodynamics and some other fields of engineering.

We want to thank the director of the Oberwolfach Institute, Prof. Barner, who gave us the opportunity to organize this meeting. We also express our thanks to Dr. Gipser and Dipl.-Math. Kaspar, who coordinated the production of the manuscript, and last but not least to the editors of the Lecture Notes series and the Springer-Verlag for publishing this volume.

Hamburg, Köln, and Darmstadt, June 1982

R. Ansorge, Th. Meis, W. Törnig

LIST OF CONTRIBUTORS

Alefeld, G. Prof. Dr.
Institut f. Angew. Mathematik
Universität Karlsruhe
Kaiserstraße 12
D-75oo Karlsruhe

Axelsson, O. Prof. Dr.
Department of Mathematics
University of Nijmegen
NL-6525 Nijmegen

Hackbusch, W. Prof. Dr.
Abteilung f. Mathematik
Universität Bochum
Universitätsstr. 15o, Geb. NA
D-463o Bochum-Querenburg

Hornung, U. Dr.
Institut f. Numerische und
Instrumentelle Mathematik
Universität Münster
Einsteinstraße 64
D-44oo Münster

Kaspar, B. Dipl.-Math.
Fachbereich Mathematik
TH Darmstadt
Schloßgartenstraße 7
D-61oo Darmstadt

Meis, Th. Prof. Dr.
Mathematisches Institut
Universität Köln
Weyertal 86 - 9o
D-5ooo Köln 41

Mittelmann, H. D. Prof. Dr.
Abteilung Mathematik
Universität Dortmund
Postfach 5oo5oo
D-46oo Dortmund 5o

Neumaier, A. Dr.
Institut f. Angew. Mathematik
Universität Freiburg
Hermann-Herder-Str. 1o
D-78oo Freiburg

Nickel, K. Prof. Dr.
Institut f. Angew. Mathematik
Universität Freiburg
Hermann-Herder-Str. 1o
D-78oo Freiburg

Potra, F. A. Prof. Dr.
Department of Mathematics
National Institute for Scientific
and Technical Creation
Bd. Pacii 22o
79622 Bukarest, Romania

Niethammer, W. Prof. Dr.
Institut f. Prakt. Mathematik
Universität Karlsruhe
Englerstraße 2
D-75oo Karlsruhe

Weiland, C. Dr.
MBB-Flugzeuge GmbH
Postfach 8o116o
D-8ooo München 8o

Dr. W. Werner
Fachbereich Mathematik
Universität Mainz
Saarstraße 21
D-65oo Mainz

CONTENTS

ON GLOBAL CONVERGENCE OF ITERATIVE METHODS

O. Axelsson

Department of Mathematics,
University of Nijmegen
The Netherlands

We review and extend results on the local convergence of the classical Newton-Kantorovich method. Then we discuss globally convergent damped and inexact Newton methods and point out advantages of using a minimal error conjugate gradient method for the linear systems arising at each Newton step.

Finally application on a nonlinear elliptic problem is considered. A combination of nested iterations, damped inexact Newton method and two-level grid finite element methods for the solution of the linear boundary value problems encountered at each step are discussed.

1. Introduction

For the solution of a nonlinear problem

$$F(\underline{x}) = \underline{0} \quad , \quad F : X \to X,$$

X a Banachspace, to which we assume that there exists a solution $\hat{\underline{x}}$, we consider methods on the form

$$(1.1) \qquad C_k(\underline{x}^{k+1} - \underline{x}^k) = -\tau_k F(\underline{x}^k).$$

Here C_k is nonsingular and in some sence makes $C_k^{-1}F(\underline{x}^k)$ locally close to the solution and approximately behave like $\underline{x}^k - \hat{\underline{x}}$. There are two main types of choices of C_k:
(i) If there exists a linear operator A such that $\|F(\underline{x}^k) - A\underline{x}^k\|$ is almost independent on $\underline{x}^k$, then we let $C_k = A$. In this case the problem is almost linear and we may use iterations of Picard type. An example is given in Section 2.
(ii) If F is Fréchet differentiable, then we may let $C_k = F'(\underline{x}_k)$ and (1.1) becomes the classical Newton-Kantorovich method (with damping parameter τ_k). We may also let C_k be an approximation of $F'(\underline{x}_k)$.

The classical Newton method suffers from two disadvantages. Firstly, it is in general only locally convergent. This is discussed in Section 3. Secondly, at each step we have to solve a linear system of equations "exactly", and this may not be justified in particular when the approximations of the nonlinear system are far from the solution.

Hence we consider inexact Newton methods where the Newton step is calculated by an iterative method and the iterations are stopped when the residual of the linear system is small enough.

The global convergence is achieved when we use damped steplengths. Under the assumption of nonsingularity of F', we prove that there exists steplengths such that convergence is achieved for all initial approximations. However we point out that it may take many steps before we can achieve the superlinear convergence, which characterizes Newton type methods when we are close to a solution. This may be improved upon by use of a minimal error iterative algorithm instead of a minimal residual algorithm for the linear systems.

Finally we discuss an application on nonlinear boundary value problems. We propose a combination of a continuation method, through finer and finer meshes, with the use of a damped Newton method to solve the nonlinear system at each grid. The solution of the linear systems encountered at each Newton step, may be solved by finite elements and a multigrid method of two-level type. If we want a discretization error $O(h^p)$ at the final mesh, with meshparameter h; the number of continuation steps will be $O(|\log h|)$. The computational complexity at each mesh is of optimal order, i.e. only proportional to the number of meshpoints. The order of the discretization error for the nonlinear problem follows directly from that valid for the corresponding linearized problems.

2. Picard iteration on a mildly nonlinear singularly perturbed boundary value problem.

Consider the boundary value problem

$$-\varepsilon \underset{\sim}{\nabla} \cdot \underset{\sim}{\nabla} u + \underset{\sim}{b} \cdot \underset{\sim}{\nabla} u + cu = f(u) \quad , \underline{x} \in \Omega \in \mathbb{R}^2$$

$$u = 0 \quad \text{on} \quad \partial\Omega.$$

We assume that $\varepsilon > 0$, $c \geq 0$ and that $\frac{\partial f}{\partial u}$ is bounded on $\bar{\Omega}$. If the coefficients and the boundary are smooth enough there exists a solution in $C^2(\Omega)$ and if $\frac{\partial f}{\partial u} \leq 0$ on Ω, it is unique.

We discretize the problem by finite differences in such a way that the discretization operator L_h satisfies

1. L_h is monotone (for instance, L_h is of positive type and is positive definite).

2. There exists a barrier function $w \geq 0$ such that $L_h w \geq \delta > 0 \quad \forall \underline{x} \in \Omega_h$.

<u>Definition 2.1.</u> Let $|u|_{\Omega_h} = \max_{\underline{x}\in\Omega_h} |u|$.

We have then the following wellknown result, often called the Barrier-Lemma. For every u on Ω_h with $u = 0$ on $\partial\Omega_h$,

$$|u|_{\Omega_h} \leq \frac{\max\limits_{\Omega_h} w}{\min\limits_{\Omega_h} L_h w} |L_h u|_{\Omega_h} . \tag{2.1}$$

For the construction of a barrier function we assume that the first component of the velocity vector $\underline{b} = (b_1, b_2)$ satisfies

$$b_1(\underline{x}) \geq b_0 > 0 \quad \forall \underline{x} \in \Omega .$$

Let

$$w(x,y) = |x|^2_\Omega - x^2 + 3R(|x|_\Omega + x)$$

where R is the radius of a circle with center at the origin and which covers Ω. We let L_h be a central difference operator with h small enough (needed for positivity) or we use central differences for the second order term but upwind for the first order terms. Then we have

$$w \geq 0$$

and

$$\begin{aligned} L_h w &= 2\varepsilon + b_1(3R - 2x + \theta h) + cw \\ &\geq 2\varepsilon + R\, b_0 \quad \forall (x,y) \in \Omega_h , \end{aligned}$$

where $\theta = \begin{cases} 0 \text{ for central difference scheme} \\ 1 \text{ for upwind difference scheme.} \end{cases}$

It follows from (2.1) that

$$|u|_{\Omega_h} \leq \frac{6R^2}{2\varepsilon + Rb_0} |L_h u|_{\Omega_h} . \tag{2.2}$$

We shall now solve the discretized equation

$$L_h u_h = f(u_h) \tag{2.3}$$

by Picard iteration,

$$L_h u_h^{(i)} = f(u_h^{(i-1)}) , \quad i = 1,2,... \tag{2.4}$$

where $u_h^{(0)}$ may be arbitrarily chosen. From (2.3), (2.4) we get

$$L_h(u_h - u_h^{(i)}) = \frac{\partial f}{\partial u} (u_h - u_h^{(i-1)})$$

and by (2.2) we then get

$$|u_h - u_h^{(i)}|_{\Omega_h} \le \frac{6R^2}{2\varepsilon+R\, b_0} \max_u |\frac{\partial f}{\partial u}(u(.))|_{\Omega_h} |u_h - u_h^{(i-1)}| .$$

Hence, if $|\frac{\partial f}{\partial u}|$ is small enough

$$|u_h - u_h^{(i)}|_{\Omega_h} \le \rho |u_h - u_h^{(i-1)}|_{\Omega_h} \quad , \; i = 1,2,\ldots$$

where $0 < \rho < 1$, and we have convergence for every $u_h^{(0)}$.
Naturally, $|\frac{\partial f}{\partial u}|$ small means that the problem is almost linear.

3. <u>Newton-Kantorivich method, local convergence.</u>

Simplified proofs of classical results on local convergence of Newton-Kantorovich method (see Kantorovich [6], Rall [10], and Ortega, Rheinboldt [9]) are presented. The results are extended.

We assume that the mapping $F : X \to X$, where X is a Banach space, is continuously Fréchet differentiable with derivative F' nonsingular for all $\underline{\xi} \in X$, with bound

$$\beta = \max_{\xi \in X} \| F'(\underline{\xi})^{-1} \| .$$

Further we assume that F' is Lipschitz continuous,

$$\text{(3.1)} \qquad \| F'(\underline{\xi}) - F'(\underline{\eta}) \| \le K \| \underline{\xi} - \underline{\eta} \| \qquad \forall \xi, \eta \in X,$$

and that there exists a solution $\hat{\underline{\xi}}$ of $F(\underline{\xi}) = 0$.
(A solution exists, and is unique, if in addition F is norm-coercive, i.e. $\lim \| F(\underline{\xi}) \| = +\infty$, $\| \underline{\xi} \| \to \infty$.) Let $\{\underline{\xi}^k\}$ be the Newton sequence

$$\text{(3.2)} \qquad \underline{\xi}^{k+1} = \underline{\xi}^k - F'(\underline{\xi}^k)^{-1} F(\underline{\xi}^k) \quad , \quad k = 0,1,\ldots$$

with $\underline{\xi}^0$ given. We have

<u>Theorem 3.1.</u>

Assume that $d = \frac{1}{2}\beta K \| \underline{\xi}^1 - \underline{\xi}^0 \| < 1$.

Then for any F satisfying the above conditions, we have

$$\| \underline{\xi}^k - \hat{\underline{\xi}} \| \le \frac{2}{\beta K} \sum_{i=k}^{\infty} d^{(2^i)} .$$

<u>Proof.</u> Using the integral relation

$$\text{(3.3)} \qquad F(\underline{\eta}) = F(\underline{\xi}) + F'(\underline{\xi})(\underline{\eta}-\underline{\xi}) + \{\int_0^1 [F'(\underline{\xi} + t(\underline{\eta}-\underline{\xi})) - F'(\underline{\xi})]dt\}(\underline{\eta}-\underline{\xi}) \; ,$$

we get

$$\|F(\underset{\sim}{\xi}^k)\| = \|F(\underset{\sim}{\xi}^{k-1}) + F'(\underset{\sim}{\xi}^{k-1})(\underset{\sim}{\xi}^k-\underset{\sim}{\xi}^{k-1}) + \{\int_0^1 [F'(\underset{\sim}{\xi}^{k-1} + t(\underset{\sim}{\xi}^k-\underset{\sim}{\xi}^{k-1})) - F'(\underset{\sim}{\xi}^{k-1})]dt\}(\underset{\sim}{\xi}^k-\underset{\sim}{\xi}^{k-1}))\| \tag{3.4}$$

and by (3.1) and (3.2),

$$\|F(\underset{\sim}{\xi}^k)\| \le \tfrac{1}{2}K\|\underset{\sim}{\xi}^k-\underset{\sim}{\xi}^{k-1}\|^2. \tag{3.5}$$

Hence by (3.2),

$$\|\underset{\sim}{\xi}^{k+1}-\underset{\sim}{\xi}\| = \|F'(\underset{\sim}{\xi}^k)^{-1}F(\underset{\sim}{\xi}^k)\| \le \tfrac{1}{2}\beta K\|\underset{\sim}{\xi}^k-\underset{\sim}{\xi}^{k-1}\|^2,$$

so

$$\tfrac{1}{2}\beta K\|\underset{\sim}{\xi}^2-\underset{\sim}{\xi}^1\| \le (\tfrac{1}{2}\beta K)^2\|\underset{\sim}{\xi}^1-\underset{\sim}{\xi}^0\|^2 = d^2$$

and by induction,

$$\tfrac{1}{2}\beta K\|\xi^{j+1}-\xi^j\| \le (\tfrac{1}{2}\beta K\|\underset{\sim}{\xi}^j-\underset{\sim}{\xi}^{j-1}\|)^2 \le (d^{(2^{j-1})})^2 = d^{(2^j)}, \quad j = 1,2,\ldots .$$

Thus

$$\|\underset{\sim}{\xi}^{k+p}-\underset{\sim}{\xi}^k\| \le \sum_{j=k}^{k+p-1}\|\underset{\sim}{\xi}^{j+1}-\underset{\sim}{\xi}^j\| \le \frac{2}{\beta K}\sum_{j=k}^{k+p-1} d^{(2^j)}, \quad k \ge 1. \tag{3.6}$$

Hence this Cauchy-sequence converges, and since X is a complete space, there exists a limit $\hat{\underset{\sim}{\xi}}$ such that

$$\|\hat{\underset{\sim}{\xi}}-\underset{\sim}{\xi}^k\| \le \frac{2}{\beta K}\sum_{j=k}^{\infty} d^{(2^i)}.$$

Further, from (3.5), if follows that

$$\|F(\underset{\sim}{\xi}^k)\| \to 0, \quad k \to \infty$$

so the limit is a solution of $F(\underset{\sim}{\xi}) = 0$. □

Clearly, this is a local convergence theorem, because in general $d < 1$ only if already $\underset{\sim}{\xi}^0$ is sufficiently close to the solution. Note also that, given upper bounds for β and K, and having calculated $\underset{\sim}{\xi}^1$, we may test the sufficiency condition $d < 1$ (for convergence of the Newton sequence) à priori.
However, it is clearly overly pessimistic to let β be the supremum of $\|F'(\underset{\sim}{\xi})^{-1}\|$ over the whole space X and in fact, in practice β may not be available. We shall now present a theorem where we only have to estimate $F'(\underset{\sim}{\xi})^{-1}$ at the initial point. The theorem, which was originally given by Kantorovich [6] (see also Rall [10]), will be presented with a simplified proof and a slight improvement. Let

$$B(\underset{\sim}{\xi},r) = \{\underset{\sim}{\eta}, \|\underset{\sim}{\eta}-\underset{\sim}{\xi}\| < r\}$$

be the open ball with center at $\underset{\sim}{\xi}$ and radius r.

<u>Theorem 3.2.</u> Let $\beta_k = \|F'(\underline{\xi}^k)^{-1}\|$, $k = 0,1,2,\ldots$. Then, if

$$h_0 := \underline{\beta_0 K\|\underline{\xi}^1-\underline{\xi}^0\|} \le \frac{1}{2+\delta}, \quad \delta \ge 0,$$ the Newton sequence (3.2) converges to a solution $\hat{\underline{\xi}} \in B(\underline{\xi}^0, r_0)$, where

$$r_0 = f(\delta)\|\underline{\xi}^1-\underline{\xi}^0\|$$

and

$$f(\delta) = 1 + d_1 + d_1 d_2 + d_1 d_2 d_3 + \ldots ,$$

$$d_k = \tfrac{1}{2}\beta_k K\|\underline{\xi}^k-\underline{\xi}^{k-1}\|, \quad k = 1,2,\ldots . \tag{3.7}$$

We have $d_k \le \frac{1}{2(1+\delta)}$ and $f(\delta) \le \frac{2(1+\delta)}{1+2\delta}$. The Lipschitz constant K satisfies

$$K = \sup_{\underline{\xi},\underline{\eta}\in B(\underline{\xi}^0,r_0)} \frac{\|F'(\underline{\xi}) - F'(\underline{\eta})\|}{\|\underline{\xi}-\underline{\eta}\|} . \tag{3.8}$$

<u>Proof:</u> From (3.5) follows

$$\begin{aligned} \|\underline{\xi}^{k+1}-\underline{\xi}^k\| &\le \tfrac{1}{2}\beta_k K\|\underline{\xi}^k-\underline{\xi}^{k-1}\|^2 \\ &\le d_k\|\underline{\xi}^k-\underline{\xi}^{k-1}\| \le\ldots\le d_k d_{k-1} \ldots d_1\|\underline{\xi}^1-\underline{\xi}^0\| . \end{aligned} \tag{3.9}$$

From

$$\begin{aligned} F'(\underline{\xi})^{-1} &= F'(\underline{\xi}^k)^{-1} + F'(\underline{\xi})^{-1} - F'(\underline{\xi}^k)^{-1} \\ &= F'(\underline{\xi}^k)^{-1} + F'(\underline{\xi})^{-1}[F'(\underline{\xi}^k) - F'(\underline{\xi})]F'(\underline{\xi}^k)^{-1} \end{aligned}$$

we find by (3.8)

$$\|F'(\underline{\xi})^{-1}\| \le \beta_k[1 + \|F'(\underline{\xi})^{-1}\|K\|\underline{\xi}^k-\underline{\xi}\|] \tag{3.10}$$

Let

$$h_k = \beta_k K\|\underline{\xi}^{k+1}-\underline{\xi}^k\| \tag{3.11}$$

Then, if $h_k < 1$, by (3.10)

$$\beta_{k+1} = \|F'(\underline{\xi}^{k+1})^{-1}\| \le \beta_k/(1-h_k) \tag{3.12}$$

and by (3.7),

$$\begin{aligned} d_k &\le \tfrac{1}{2}\beta_{k-1}K\|\underline{\xi}^k-\underline{\xi}^{k-1}\|/(1-h_{k-1}) \\ &\le \tfrac{1}{2}h_{k-1}/(1-h_{k-1}) . \end{aligned} \tag{3.13}$$

Further, by (3.11), (3.9), (3.12) and (3.13),

$$\begin{aligned} h_k &\le \beta_k K\, d_k\|\underline{\xi}^k-\underline{\xi}^{k-1}\| \le \frac{1}{1-h_{k-1}}(\beta_{k-1}K\|\underline{\xi}^k-\underline{\xi}^{k-1}\|)d_k \\ &= \tfrac{1}{2}\left(\frac{h_{k-1}}{1-h_{k-1}}\right)^2 . \end{aligned} \tag{3.14}$$

Hence, if the sequence $\{h_k\}$ shall stay bounded, in general we must have

$$h_k \le h \quad , \quad k = 1,2,\ldots$$

where h is a solution of $h = \frac{1}{2}\left(\frac{h}{1-h}\right)^2$.

The only solution satisfying $0 < h < 1$ is $h = \frac{1}{2}$. With $h_0 = \frac{1}{2+\delta}$, $\delta \geq 0$ we get by induction from (3.14)

$$h_k \leq \tfrac{1}{2}\left(\frac{1}{1+\delta}\right)^2 \leq \frac{1}{2+\delta}$$

and by (3.13)

$$d_k \leq \frac{1}{2(1+\delta)} \leq \tfrac{1}{2}.$$

Hence by (3.5), (3.9),

$$\| F(\underline{\xi}^k) \| \to 0 \ , \ k \to \infty.$$

By (3.9) we have

$$\| \underline{\xi}^p - \underline{\xi}^0 \| \leq \sum_{j=0}^{p-1} \| \underline{\xi}^{j+1} - \underline{\xi}^j \| \leq f(\delta) \| \underline{\xi}^1 - \underline{\xi}^0 \|$$

so

$$\| \underline{\xi} - \underline{\xi}^0 \| \leq f(\delta) \| \underline{\xi}^1 - \underline{\xi}^0 \|$$

and

$$f(\delta) \leq \sum_{j=0}^{\infty} \left(\frac{1}{2(1+\delta)}\right)^\delta = \frac{2(1+\delta)}{1+2\delta} .$$

This proves the theorem. □

By a more accurate estimate we have by (3.14), (3.13)

$$\left.\begin{aligned} h_0 &= \frac{1}{2+\delta} , \ \delta > 0 \\ d_k &:= \tfrac{1}{2} \frac{h_{k-1}}{1-h_{k-1}} , \\ h_k &:= \tfrac{1}{2}\left(\frac{h_{k-1}}{1-h_{k-1}}\right)^2 \\ \rho_k &:= \rho_{k-1} + \prod_{j=1}^{k} d_j \end{aligned}\right\} \quad k = 1,2,\ldots$$

Then

$$r_0 \leq \rho(\delta) \| \underline{\xi}^1 - \underline{\xi}^0 \| ,$$

where

$$\rho(\delta) = \lim_{k\to\infty} \rho_k$$

It is easy to see that

$$\rho(\delta) \leq \frac{2}{1 + \sqrt{1-2h_0}} ,$$

the latter being the bound given in [6], [10].

Note that Theorem 3.2 gives an à priori localization result for the solution. In practical applications, the following improved version of Theorem 3.1 may be of greater importance, in particular in cases where β_0 is large.

<u>Theorem 3.3.</u> Let the <u>relative Lipschitz</u> constants,

$$\tilde{K}_k = \sup_k 2 \frac{\| \int_0^1 F'(\underline{\xi}^k)^{-1}[F'(\underline{\xi}^{k-1} + t(\underline{\xi}^k-\underline{\xi}^{k-1})) - F'(\underline{\xi}^{k-1})]dt \|}{\| \underline{\xi}^k-\underline{\xi}^{k-1} \|}$$

and

$$\tilde{K} = \max_{k\geq 1} \tilde{K}_k$$

If $\tilde{d} = \frac{1}{2}\tilde{K}\| \underline{\xi}^1-\underline{\xi}^0 \| < 1$, then the Newton sequence (3.2) converges to a solution $\hat{\underline{\xi}}$ and

$$\| \underline{\xi}^k-\hat{\underline{\xi}} \| \leq \frac{2}{\tilde{K}} \sum_{i=k}^{\infty} (\tilde{d})^{(2^i)} .$$

<u>Proof.</u> We have

$$\underline{\xi}^{k+1}-\underline{\xi}^k = F'(\underline{\xi}^k)^{-1}F(\underline{\xi}^k) = F'(\xi^k)^{-1}[F(\underline{\xi}^{k-1}) + F'(\underline{\xi}^{k-1})(\underline{\xi}^k-\underline{\xi}^{k-1})]$$
$$+ F'(\underline{\xi}^k)^{-1}\{\int_0^1[F'(\underline{\xi}^{k-1} + t(\underline{\xi}^k-\underline{\xi}^{k-1})) - F'(\underline{\xi}^{k-1})]dt\}(\underline{\xi}^k-\underline{\xi}^{k-1})$$

so

$$\| \underline{\xi}^{k+1}-\underline{\xi}^k \| \leq \tfrac{1}{2}\tilde{K}_k \| \underline{\xi}^k-\underline{\xi}^{k-1} \|^2$$

or

$$\tfrac{1}{2}\tilde{K}\| \underline{\xi}^{k+1}-\underline{\xi}^k \| \leq (\tfrac{1}{2}\tilde{K}\| \xi^k-\xi^{k-1} \|)^2 , \quad k = 1,2,\dots .$$

The remaining part of the proof now follows that of Theorem 3.1. □

Note that

$$\tilde{K} \leq \sup_{\underline{\xi},\underline{\eta}\in B(\hat{\xi},r)} 2 \frac{\| \int_0^1 F'(\underline{\eta})^{-1}[F'(\underline{\xi} + t(\underline{\eta}-\underline{\xi})) - F'(\xi)]dt \|}{\| \underline{\eta} - \underline{\xi} \|}$$

and hence (compare Theorem 3.1)

$$\tilde{K} \leq \beta K.$$

In practice $\tilde{K}$ may be small (or at least close to one) whereas β (and K) may be very large. Hence $\tilde{d}$ may be much smaller than d and the actual local radius of convergence may be quite large.

4. <u>On globally convergent preconditioned damped inexact Newton methods.</u>

This section is an extension of works by Dennis and Moré [5], Dembo, Eisenstat and Steihaug [4], and by Rank and Rose [3], and with modified proofs.

We note first that at each Newton step we have to calculate a step $\underline{p}^k$ from

$$F'(\underline{\xi}^k)\underline{p}^k = -F(\underline{\xi}^k) \tag{4.1}$$

and then update the approximation,

$$\underline{\xi}^{k+1} = \underline{\xi}^k + \underline{p}^k , \quad k = 0,1,\dots .$$

Computing the exact solution of (4.1) can be expensive. This will be the case for instance if we are to solve a nonlinear partial differential equation problem. Furthermore, it may not be justified when $\underline{\xi}^k$ is far from a solution. Thus one might prefer to compute an approximate solution. This can be done in the following way. Find $\underline{p}^k$ such that

(4.2) $\qquad \| M_k[F(\underline{\xi}^k) + F'(\underline{\xi}^k)\underline{p}^k] \| \le \rho_k \| M_k F(\underline{\xi}^k) \| \quad , \rho_k \in (0,1)$

and let

$$\underline{\xi}^{k+1} = \underline{\xi}^k + \underline{p}^k .$$

Here M_k is a given nonsingular operator chosen such that the condition number of $M_k F'(\underline{\xi}^k)$ is much smaller than that of $F'(\underline{\xi}^k)$. M_k is called a preconditioning operator. (In practice one often lets $N_K = M_K^{-1}$ be given and as an approximation of $F'(\underline{\xi}_k)$.)
Such a $\underline{p}^k$ obviously exists if $F'(\underline{\xi}^k)$ is nonsingular. The forcing sequence $\{\rho_k\}$ controls how accurately the Newton equations are solved, but the method of actually computing such a vector is not specified. (Needless to say, $\underline{p}^k$ is not uniqually defined.) We are thinking of using iterative methods, such as the generalized minimum residual or minimum error conjugate gradient method (see [1]). In the following we are going to choose M_k constant during iterations $k = 0,1,\ldots,k_0$, say, then possibly update M_k and use this during a number of iteration steps etc. Hence in the analysis to follow we may consider solving the problem

(4.3) $\qquad \tilde{F}(\underline{\xi}) = M_0 F(\underline{\xi}) = 0, \underline{\xi} \in X.$

In particular, we are thinking of having $M_0 = F'(\underline{\xi}^0)^{-1}$.

Another disadvantage with the Newton method is, as we have seen in Section 3, that it is in general only locally convergent, i.e. the sequence $\{\underline{\xi}_k\}$ need not converge unless $\underline{\xi}^0$ is sufficiently close to a solution. The wellknown remedy is to update with damped steps, i.e.

(4.4) $\qquad \underline{\xi}^{k+1} = \underline{\xi}^k + \tau_k \underline{p}^k \quad , \quad \tau_k \in (0,1].$

We shall now analyse the combined method, the damped inexact Newton (DIN) method. We assume that F is normcoercive, we let $\underline{\xi}^0 \in X$ be given and let

$$S_0 = \{\underline{\xi} \; ; \; \| \tilde{F}(\underline{\xi}) \| \le \| \tilde{F}(\underline{\xi}^0) \| \}.$$

We assume further that F is Fréchet differentiable and that $\tilde{F}'$ is Hölder-continuous, i.e. it satisfies

(4.5) $\qquad \| \tilde{F}'(\underline{\xi}) - \tilde{F}'(\underline{\eta}) \| \le \Gamma \| \underline{\xi} - \underline{\eta} \|^{\gamma} \; , \; \gamma \in (0,1] \quad \forall \underline{\xi}, \underline{\eta} \in S_1,$

where S_1 is the convex hull of S_0.
We also assume that F' is nonsingular on S_0.

Lemma 4.1. Given $\delta \in (0,1)$, let $\{\rho_k\} \in (0,1-\delta)$ be a forcing sequence in (4.2) with $M_k = M_0$, $k = 0,1,\ldots$ and let the sequence $\{\underline{\xi}^k\}$ be such that (4.2), (4.4) are satisfied. Let

$$(4.6) \qquad \Gamma_k = \| \{\int_0^1 [\tilde{F}'(\underline{\xi}^k + t\tau_k \underline{p}^k) - \tilde{F}'(\underline{\xi}^k)]dt\}\underline{p}^k \| / (\tau_k^\gamma \| \tilde{F}(\underline{\xi}^k) \|^{1+\gamma}).$$

Then

$$(4.7) \qquad \Gamma_k \le \frac{\Gamma}{1+\gamma} [(1+\rho_k)\beta_k]^{1+\gamma} \le c_0 \quad , \quad k = 0,1,\ldots$$

where

$$\beta_k = \| \tilde{F}'(\underline{\xi}^k)^{-1} \|$$

and with

$$(4.8) \qquad 0 < \tau_k \le \min\{ (\frac{1-\delta-\rho_k}{\Gamma_k})^{1/\gamma} / \| \tilde{F}(\underline{\xi}^k) \| , 1\}$$

we have

$$(4.9) \qquad \| \tilde{F}(\underline{\xi}^{k+1}) \| \le (1 - \delta\tau_k) \| \tilde{F}(\underline{\xi}^k) \| , \quad k = 0,1,\ldots$$

Proof. From the integral relation (3.3) follows that

$$\begin{aligned}
\tilde{F}(\underline{\xi}^{k+1}) &= \tilde{F}(\underline{\xi}^k) + \tilde{F}'(\underline{\xi}^k)(\underline{\xi}^{k+1} - \underline{\xi}^k) \\
&+ \{\int_0^1 [\tilde{F}'(\underline{\xi}^k + t(\underline{\xi}^{k+1} - \underline{\xi}^k)) - \tilde{F}'(\underline{\xi}^k)]dt\}(\underline{\xi}^{k+1} - \underline{\xi}^k) \\
&= (1-\tau_k)\tilde{F}(\underline{\xi}^k) + \tau_k[\tilde{F}(\underline{\xi}^k) + \tilde{F}'(\underline{\xi}^k)\underline{p}^k] \\
&+ \tau_k\{\int_0^1 [\tilde{F}'(\underline{\xi}^k + t\tau_k \underline{p}^k) - \tilde{F}'(\underline{\xi}^k)]dt\}\underline{p}^k .
\end{aligned}$$

Hence by (4.2) and (4.6),

$$\| \tilde{F}(\underline{\xi}^{k+1}) \| \le e_k \| \tilde{F}(\underline{\xi}^k) \|$$

where

$$e_k = 1 - \tau_k + \tau_k\rho_k + \tau_k^{1+\gamma} \Gamma_k \| \tilde{F}(\underline{\xi}^k) \|^\gamma .$$

We have

$$\frac{1}{\tau_k}[1-\delta\tau_k - e_k] = 1-\delta-\rho_k - \tau_k^\gamma \Gamma_k \| \tilde{F}(\underline{\xi}^k) \|^\gamma \ge 0,$$

by (4.8). Hence $e_k \le 1-\delta\tau_k$ and (4.9) follows.
In particular, by induction, $\{\underline{\xi}^k\} \in S_0$.
It remains to prove that $\{\Gamma_k\}$ is bounded. By (4.2) it follows that

$$(4.10) \qquad \| \tilde{F}'(\underline{\xi}^k)\underline{p}^k \| \le (1+\rho_k) \| \tilde{F}(\underline{\xi}^k) \|$$

or

$$\| \underline{p}^k \| \le \beta_k (1+\rho_k) \| \tilde{F}(\underline{\xi}^k) \| .$$

This together with the Hölder continuity proves (4.7). □

We may let

(4.11) $$C_0 = \frac{\Gamma}{1+\gamma}(2\beta)^{1+\gamma}$$

where

(4.12) $$\beta = \sup_{\underline{\xi}\in S_0} \|\tilde{F}'(\underline{\xi})^{-1}\| .$$

<u>Theorem 4.1.</u> (Global convergence) Given the assumptions in Lemma 4.1, let ρ_k be a nonincreasing sequence $0 < \rho_k \leq \rho_0 < 1-\delta$. Then

$$\|\tilde{F}(\underline{\xi}^k)\| \leq (1-\delta\tau)^k \|\tilde{F}(\underline{\xi}^0)\|$$

where

$$\tau_k \geq \tau = \min\{(\frac{1-\delta-\rho_0}{C_0})^{1/\gamma} / \|\tilde{F}(\underline{\xi}^0)\| ,1\}$$

and C_0 is defined in (4.11).

<u>Proof.</u> By Lemma 4.1,

$$\|\tilde{F}(\underline{\xi}^k)\| \leq (1-\delta\tau_k)\|\tilde{F}(\underline{\xi}^{k-1})\| \leq (1-\delta\tau)\|\tilde{F}(\underline{\xi}^{k-1})\|$$
$$\leq \ldots \leq (1-\delta\tau)^k \|\tilde{F}(\underline{\xi}^0)\| .$$ □

This theorem shows that we get global convergence from any point $\underline{\xi}^0$ such that the assumptions are satisfied on the corresponding set S_0. Naturally the convergence may be very slow, i.e. we may be forced to take very small steps $\tau_k \underline{p}^k$ which will be the case of Γ and/or β is large. To some extent, a good choice of preconditioning operator M_0 may help in this respect. Anyway, if a plateau of very slow rate of convergence is reached, it is suggested to perturb the problem by a linear term with a positive definite operator, and then gradually let the perturbation parameter go to zero (i.e. we propose a singular perturbation embedding).

We note also from (4.8), that as $\|F(\underline{\xi}^k)\|$ decreases monotonically, it follows that we may let τ_k eventually approach 1.

In order to achieve a superlinear rate of convergence we shall now in fact choose the sequence $\{\tau_k\}$ such that it automatically approaches 1 as $k \to \infty$ and furthermore, the sequence $\{\rho_k\}$ such that ρ_k converges towards zero fast enough.

<u>Theorem 4.2.</u>

Let in the DIN algorithm

(4.13) $$\tau_k = \frac{1}{1+s_k\|\tilde{F}(\underline{\xi}^k)\|^q}$$

and

(4.14) $$\rho_k = c_0\|\tilde{F}(\underline{\xi}^k)\|^q, \quad q \in (0,1].$$

Here $s_0 \geq s_k \geq 0$ and is large enough so that

$$\tau_k \leq \left(\frac{1-\delta-\rho_0}{c_0}\right)^{1/\gamma} / \|\tilde{F}(\underline{\xi}^k)\| \quad , \ k = 0,1,\ldots$$

is satisfied. Further, let

$$0 < c_0 \leq (1-\delta)/\|\tilde{F}(\underline{\xi}^0)\|^q$$

Then

$$\|\hat{\underline{\xi}} - \underline{\xi}^{k+1}\| \leq c\|\hat{\underline{\xi}} - \underline{\xi}^k\|^{1+\min(\gamma,q)}$$

Proof. We have by the integral relation (3.3),

$$0 = \tilde{F}(\hat{\underline{\xi}}) = \tilde{F}(\underline{\xi}^k) + \tilde{F}'(\underline{\xi}^k)(\hat{\underline{\xi}} - \underline{\xi}^k) +$$
$$\{\int_0^1[\tilde{F}'(\underline{\xi}^k + t(\hat{\underline{\xi}} - \underline{\xi}^k)) - \tilde{F}'(\underline{\xi}^k)]dt\}(\hat{\underline{\xi}} - \underline{\xi}^k).$$

Hence by (4.4)

$$0 = \tilde{F}(\underline{\xi}^k) + \tau_k \tilde{F}'(\underline{\xi}^k)\underline{p}_k + \tilde{F}'(\underline{\xi}^k)(\hat{\underline{\xi}} - \underline{\xi}^{k+1}) +$$
$$\{\int_0^1[\tilde{F}'(\underline{\xi}^k + t(\hat{\underline{\xi}} - \underline{\xi}^k)) - \tilde{F}'(\underline{\xi}^k)]dt\}(\hat{\underline{\xi}} - \underline{\xi}^k)$$

so by (4.12), (4.2) and (4.5),

$$\frac{1}{\beta}\|\hat{\underline{\xi}} - \underline{\xi}^{k+1}\| \leq \tau_k\rho_k\|\tilde{F}(\underline{\xi}^k)\| + (1-\tau_k)\|\tilde{F}(\underline{\xi}^k)\| + \frac{\Gamma}{1+\gamma}\|\hat{\underline{\xi}} - \underline{\xi}^k\|^{1+\gamma}.$$

But

$$\|\tilde{F}(\underline{\xi}^k)\| = \|\tilde{F}(\underline{\xi}^k) - \tilde{F}(\hat{\underline{\xi}})\| \leq C_1\|\underline{\xi}^k - \hat{\underline{\xi}}\|$$

where

$$C_1 = \sup_{\xi \in S_1} \|\tilde{F}'(\xi)\|.$$

Hence by (4.10), (4.13) and (4.14)

$$\|\hat{\underline{\xi}} - \underline{\xi}^{k+1}\| \leq \beta\{[c_0 + s_0]\|\tilde{F}(\underline{\xi}^k)\|^{1+q} + \frac{\Gamma}{1+\gamma}\|\hat{\underline{\xi}} - \underline{\xi}^k\|^{1+\gamma}\}$$
$$\leq \beta\{[c_0 + s_0]C_1^{1+q} + \frac{\Gamma}{1+\gamma}\}\|\hat{\underline{\xi}} - \underline{\xi}^k\|^{1+\min(\gamma,q)}$$

□

In practice we let $q \geq \gamma$ and it follows then from this theorem and the previous Lemma that eventually the DIN sequence converges superlinearly, with convergence of order $(1+\gamma)$. If γ is not known, then we choose $q = 1$.

5. Minimal error conjugate gradient algorithm

As an alternative to the (preconditioned) conjugate gradient DIN algorithm

(4.2) we may use a minimal error algorithm, i.e. we may calculate a step direction $\underline{p}^k$ such that

(5.1) $$\| F'(\underline{\xi}^k)^{-1}[F(\underline{\xi}^k) + F'(\underline{\xi}^k)\underline{p}^k]\| \leq \rho_k \| F'(\underline{\xi}^k)^{-1}F(\underline{\xi}^k)\| .$$

Clearly all corresponding constants, such as β, Γ_k and C_0 in (4.12), (4.7) and (4.11) become closer to one when $\tilde{F}(\underline{\xi}) := F'(\underline{\xi})^{-1}F(\underline{\xi})$, and the superlinear rate of convergence will begin at an earlier stage than when we apply a minimum residual algorithm, because we may now take longer steps in the beginning (see (4.8)). As is wellknown (see for instance [1], and the references quoted therein) a minimal error algorithm works in the following way.

In order to solve a system

$$A\hat{\underline{x}} = \underline{b} \quad , \quad A : X \to X \quad , \quad \underline{b} \in X,$$

we solve at first

$$A\,A^T\underline{y} = \underline{b}$$

by the classical conjugate gradient method. When the solution (or an approximation thereof) is found, we let $\hat{\underline{x}} = A^T\hat{\underline{y}}$. At the m'th step of this algorithm we calculate

$$\min_{\underline{y}\in S_m} f(\underline{y}) = \min_{\underline{y}\in S_m} \tfrac{1}{2}(\underline{y}-\hat{\underline{y}})^T AA^T(\underline{y}-\hat{\underline{y}}) = \min_{\underline{x}\in A^T S_m} \tfrac{1}{2}(\underline{x}-\hat{\underline{x}})^T(\underline{x}-\hat{\underline{x}})$$

which means that we minimize the <u>error</u> on $A^T S_m$.

Here

$$S_m = \{\underline{y}^0\} \cup \text{SPAN}\{\underline{g}^0, AA^T\underline{g}^0, \ldots, (AA^T)^m\underline{g}^0\},$$

$\underline{g}^0 = AA^T\underline{y}^0 - \underline{b}$ and $\underline{y}^0$ is an initial approximation. It follows that if $f(\underline{y}^m) \leq \tfrac{1}{2}\varepsilon_k^2$, then

$$\| \underline{x}^m - \hat{\underline{x}}\| \leq \varepsilon_k .$$

Applying this algorithm on $F'(\underline{\xi}^k)\underline{p}^k + F(\underline{\xi}^k) = 0$ means that the error at the m'th (inner) step

$$\| F'(\underline{\xi}^k)^{-1}F(\underline{\xi}^k) + \underline{p}^{k,m}\|$$

will be smaller than any $\varepsilon > 0$, if only m is large enough. The problem with the algorithm is that we don't know when to break, because

$$\varepsilon_k = \rho_k \| F'(\underline{\xi}^k)^{-1}F(\xi^k)\| ,$$

a number which is clearly not available. We note however that the difference

$$f(\underline{y}^m) - f(\underline{y}^{m+1}) = \tfrac{1}{2}(\| A^T\underline{y}^m\|^2 - \| A^T\underline{y}^{m+1}\|^2) - (\underline{y}^m - \underline{y}^{m+1})^T\underline{b} ,$$

is available. Hence we may use a test such as

(5.2) $$f(\underline{y}^m) - f(\underline{y}^{m+1}) \leq \tfrac{1}{2}\eta_k^2\| \underline{x}^m\|^2 .$$

When this is satisfied , we have

$$\| \underline{x}^m-\hat{\underline{x}}\|^2 - \| \underline{x}^{m+1}-\hat{\underline{x}}\|^2 \le \eta_k^2 \| \underline{x}^m \|^2.$$

If η_k^2 is small, this means that either the rate of convergence is very slow, (i.e. $\| \underline{x}^m-\hat{x}\| \simeq \| \underline{x}^{m+1}-\hat{\underline{x}}\|$) or that we are close to a solution (i.e. $\| \underline{x}^m-\hat{\underline{x}}\| \simeq \eta_k^2 \| \underline{x}^m \|^2$).

Now, if we use a good preconditioning operator M_k, the convergence will not be too slow and we have then, if (5.2) is satisfied,

(5.3) $$\| \underline{x}^m-\hat{\underline{x}}\|^2 \le \zeta\, \eta_k^2 \| \underline{x}^m \|^2$$

where, say, $0 < \zeta \le 4$. Hence

$$\| \underline{x}^m-\hat{\underline{x}}\| \le \sqrt{\zeta}\, \eta_k \| \underline{x}^m \|$$

and we may choose, say, $\eta_k = \rho_k/2$.

(If the norm of the solution happens to be small, it may be advisible to add a constant to the solution and proceed to solve the transformed problem.)

In many cases, we may get an estimate of the number ζ in (5.3) from the sequence $\{f(\underline{y}^m) - f(\underline{y}^{m+1})\}_{m\ge 0}$.

6. Application on nonlinear boundary value problems.

We consider now a nonlinear boundary value problem of 2m'th order on a variational form; Find $\tilde{u}$ such that

(6.1) $$(F(u),\eta) = 0 \quad \forall \eta \in V,$$

where $V = H_0^m(\Omega)$, $F(u) \in V^*$, the dual space and $u = \tilde{u}+\gamma$, $\tilde{u} \in V$ and $\gamma \in H^m(\Omega)$ arbitrary but the traces of γ satisfies the given essential boundary conditions on $\partial\Omega$. We assume that F' is Gateaux differentiable and that its differential F' is such that the bilinear form

$$a(u\ ;\ \eta_1,\eta_2) = (F'(u)\eta_1,\eta_2)\ ,\ \eta_1,\eta_2 \in V$$

is symmetric and coercive, i.e.

(6.2) $$a(u\ ;\ \eta,\eta) \ge \alpha \| \eta \|_V^2 \quad \forall \eta \in V \quad \forall \tilde{u} \in V,\ \alpha > 0$$

and bounded, i.e.

(6.3) $$a(u\ ;\ \eta_1,\eta_2) \le \beta \| \eta_1 \|_V \| \eta_2 \|_V \quad \forall \eta_1,\eta_2 \in V,\ \tilde{u} \in V.$$

Further we assume that F' is Hölder continuous on V.[We remark that it suffices that the constants α, β, Γ and similar constants to follow, are valid uniformly in u on a set corresponding to S_0 in Section 4.]

We let $\{\Omega_{h_i}\}_{i\ge 0}$ be sequence of admissible meshes with $h_0 > h_1 > h_2 > \ldots$ and

$h_i \to 0$, $i \to \infty$. This means that $\Omega_{h_i} \subset \Omega_{h_{i+1}}$ and the vertices in Ω_{h_i} form a subset of the vertices in $\Omega_{h_{i+1}}$. We also assume that $\overline{\Omega}_{h_i} = \overline{\Omega}$, which means that the boundary $\partial\Omega$ is defined by piecewise and continuous polynomial functions of the same degree as the finite element basis functions $\{\phi_j^{(h_i)}\}$ to be used on Ω_{h_i}, $i = 0,1,\ldots$.

We shall prove that algorithm (4.2) gives us a framework in order to construct a sequence of functions u_{h_i} on Ω_{h_i}, $i = 0,1,2,\ldots$ that converges to the exact solution of (6.1). Naturally by $u_{h_i} = \sum_j \alpha_j^{(h_i)} \phi_j^{(h_i)}$, this function is defined on the whole of Ω.

Given u_{h_0}, at step k of the algorithm we want to determine an approximate solution $u_{h_{k+1}}$ such that

$$\|F(u_{h_{k+1}})\|_{V^*} \le \rho_k \|F(u_{h_k})\|_{V^*}, \quad 0 < \rho_k < 1, \quad k = 0,1,2,\ldots . \tag{6.4}$$

However, we can only test if

$$\|F(u_{h_{k+1}})\|_{V^*_{h_{k+1}}} \le \zeta_k \rho_k \|F(u_{h_k})\|_{V^*_{h_{k+1}}}, \tag{6.5}$$

where

$$\|F(u)\|_{V^*_{h_{k+1}}} = \sup_{v \in V_{h_{k+1}}} \frac{(F(u),v)}{\|v\|_V},$$

i.e. a norm in a finite dimensional subspace of V. Our aim is to show that ζ_k may be chosen small enough so that (6.4) is satisfied. We have

$$\|F(u_{h_{k+1}})\|_{V^*} = \|F(u_{h_k}) + A_k(u_{h_{k+1}} - u_{h_k}\|_{V^*}$$

where

$$A_k = \int_0^1 F'(u_{h_k} + t(u_{h_{k+1}} - u_{h_k}))dt,$$

a mapping $V \to V^*$. We let $\hat{u}_{k+1} \in V$ be the exact solution of

$$F(u_{h_k}) + A_k(\hat{u}_{k+1} - u_{h_k}) = 0. \tag{6.6}$$

Hence

$$\begin{aligned} \|F(u_{h_{k+1}})\|_{V^*} &= \|A_k(u_{h_{k+1}} - \hat{u}_{k+1}\|_{V^*} \\ &\le \max_{v \in V}\|F'(v)\|_{V^*} \|u_{h_{k+1}} - \hat{u}_{h_{k+1}}\|_V + D_k \end{aligned} \tag{6.7}$$

where $\hat{u}_{h_{k+1}} \in V_{h_{k+1}}$ is the Galerkin solution of problem (6.6), i.e. the "elliptic" projection of $\hat{u}_{k+1}$ onto $V_{h_{k+1}}$ and where

$$D_k = \|A_k(\hat{u}_{h_{k+1}} - \hat{u}_{k+1})\|_{V^*}.$$

Note that $\hat{u}_{h_{k+1}} = u_{h_{k+1}}$ iff $\|F(u_{h_{k+1}})\|_{V_{h_{k+1}}} = 0$.

Let

$$(6.8) \qquad C_k = \|F(u_{h_k})\|_{V^*}.$$

Note that if $C_k = 0$ then already u_{h_k} is the exact solution. Hence we may assume that $C_k > 0$, $k = 0,1,\ldots$. Let

$$(6.9) \qquad \delta_k = 1 - D_k/C_{k+1} .$$

Since $\hat{u}_{h_{k+1}}$ is the elliptic projection, $D_k < C_{k+1}$ so

$$(6.10) \qquad 0 < \delta_k \le 1.$$

We have

$$\|\{\int_0^1 F'(\hat{u}_{h_{k+1}} + t(u_{h_{k+1}} - \hat{u}_{h_{k+1}}))dt\}(u_{h_{k+1}} - \hat{u}_{h_{k+1}})^2\|_{V^*_{h_{k+1}}}$$

$$= \|F(u_{h_{k+1}})(u_{h_{k+1}} - \hat{u}_{h_{k+1}})\|_{V^*_{h_{k+1}}} \le$$

$$\le \|F(u_{h_{k+1}})\|_{V^*_{h_{k+1}}} \|u_{h_{k+1}} - \hat{u}_{h_{k+1}}\|_V.$$

Hence by (6.2) and (6.5),

$$\alpha\|u_{h_{k+1}} - \hat{u}_{h_{k+1}}\|_V^2 \le \zeta_k\, \rho_k \|F(u_{h_k})\|_{V^*} \|u_{h_{k+1}} - \hat{u}_{h_{k+1}}\|_V$$

or

$$\|u_{h_{k+1}} - \hat{u}_{h_{k+1}}\|_V \le \alpha^{-1}\, \zeta_k\, \rho_k\, C_k.$$

From (6.7) and (6.9) it now follows

$$\|F(u_{h_{k+1}})\|_{V^*} \le \sup_{v\in V} \|F'(v)\|_{V^*}\, \alpha^{-1}\, \zeta_k\, \rho_k\, C_k + D_k$$

$$\le \delta_k\, \rho_k\, C_k + (1-\delta_k)C_{k+1},$$

if

$$\zeta_k \le \alpha\, \delta_k \,/\, \sup_{v\in V} \|F'(v)\|_{V^*}.$$

Hence by (6.8), (6.10),

$$\|F(u_{h_{k+1}})\|_{V^*} \le \rho_k\, C_k = \rho_k \|F(u_{h_k})\|_{V^*},$$

and (6.4) is satisfied.

Let $a_k(u,v)$ be the bilinear form defined by the operator A_k. By coercivity, we have

$$(6.11) \qquad \|\hat{u} - u_{h_{k+1}}\|_V \le \alpha^{-1} \sup_{v\in V} \frac{a_k(\hat{u} - u_{h_{k+1}}, v)}{\|v\|_V}$$

$$= \alpha^{-1}\|F(u_{h_{k+1}})\|_{V^*} .$$

By (6.7), $\| F(u_{h_{k+1}}) \|_{V^*}$ is bounded by a term which tells us how accurate we solve the equation $F(u_{h_{k+1}}) = 0$ in $V_{h_{k+1}}$ and the term D_k, which is the Galerkin discretization error of the linear problem (6.6). Assume this to be $O(h^p)$ in $\| \cdot \|_V$. Then by choosing $\rho_k = O(h_{k+1}/h_k)^p$ in (6.4), it follows by (6.11) that the residuals and the errors in $\| \cdot \|_V$ are reduced at the same order as the Galerkin discretization errors for the corresponding linear problems. Hence we do not have to carry out a separate derivation of the discretization errors for the nonlinear problems.
The above procedure of calculating a sequence of approximations $\{u_{h_k}\}$ on the nested meshes $\{\Omega_{h_k}\}$ may be looked upon as a continuation method with h as the continuation parameter.

For the calculation of a function $u_{h_{k+1}}$ satisfying (6.5), we propose the DIN algorithm of sections 4, 5. The convergence of this follows from Theorem 4.1. Note that the number of iteration steps are independent of h if we let ζ_k, ρ_k be independent on h. Since typically h_{k+1}/h_k has a fixed ratio, say ½, ρ_k and ζ_k may in fact be chosen independent on h (or k). Note also that a good preconditioning operator M_k in (4.2) at mesh $\Omega_{h_{k+1}}$ may be chosen on the basis of information available mesh Ω_{h_k}.

It remains to discuss the solution of the linear systems encountered at each Newton step, or rather, the calculation of a step $\underline{p}^k$ satisfying (4.2). For these we propose the use of a multigrid method of two-level type. In particular, the use of quadratic basis functions at midedge points and linear basis functions at vertex points has many advantages from a computational complexity point of view. For a discussion of such methods, see [2] and the references quoted therein. Here we shall only shortly survey the main ideas.

Given the bilinear form

$$a(u\,;\,v_1,v_2) = (F'(u)v_1,v_2)$$

on $V \times V$ we consider its restriction to mesh $\Omega_{h_{k+1}}$. We let U_{k+1} and W_{k+1} be subspaces of V, spanned by the linear and quadratic basis functions, respectively. Letting the initial approximation of the step $\underline{p}^{k,l}$ be $\underline{0}$, say, we calculate at step l of the DIN algorithm a sequence of approximations in the following way:

<u>Step (j+½)</u>: Solve for the Galerkin approximation

$$a(u_{h_{k+1}}^{(l)}\,;\,p_{h_{k+1}}^{k,l,j+\frac{1}{2}},v) = a(u_{h_{k+1}}^{(l)}\,;\,\hat{p}_{h_{k+1}},v)$$
$$= -(F(u_{h_{k+1}}^{(l)}),v) \qquad \forall v \in W_{k+1}$$

where

$$p_{h_{k+1}}^{k,l,j+\frac{1}{2}} = p_{h_{k+1}}^{k,l,j} + \tilde{w}\,,\quad \tilde{w} \in W_{k+1}$$

Step (j+1): Solve for the Galerkin approximation

$$a(u^{(1)}_{h_{k+1}}\,;\,p^{k,1,j+1}_{h_{k+1}},v) = a(u^{(1)}_{h_{k+1}}\,;\,\hat{p}_{h_{k+1}},v) \qquad \forall v \in U_{k+1},$$

$$p^{k,1,j+1}_{h_{k+1}} = p^{k,1,j+\frac{1}{2}}_{h_{k+1}} + \tilde{u}\ ,\ \tilde{u} \in U_{k+1},$$

$j = 0,1,\ldots,$ until the correction to the step is small enough.

One sweep ($j \to j+1$) of this gives a reduction of the residual of the linear problem by a factor γ^2, where arccos γ is the angle between the two subspaces. Typically γ^2 is about $\frac{1}{2}$. The linear systems at steps ($j+\frac{1}{2}$) (projections onto W_{k+1}) may be solved by splitting the corresponding matrix in two parts, where the first one corresponds to the use of material coefficients in the boundary value problem which are piecewise constant over elements. This part may be solved fast simply by first ordering the unknowns in a special way as for serendipity elements or by "static condensation", see [2]. The whole matrix problem is then solved fast by some simple iterative scheme and the computational complexity is $O(N(h_{k+1}))$, where $N(h_{k+1})$ is the number of node points on mesh $\Omega_{h_{k+1}}$.

The matrix corresponding to projections on the subspace U_{k+1} is an M-matrix (i.e. a "finite difference matrix". For this we may use many techniques, such as classical multigrid methods. This means that we at first damp the fast oscillatory components of the iteration error on the fine grid and then solve for a correction on a coarse grid.The whole process may be applied recursively. Then it is wellknown that we will achieve a computational complexity of optimal order. Note that if the ratio $h_{k+1}/h_k = \frac{1}{2}$, then the mesh points used for U_{k+1} are identical to the mesh points on Ω_{h_k}.

As described above, the whole process of solving the nonlinear boundary value problem on mesh $\Omega_{h_{k+1}}$ will be of optimal order, i.e. $O(N(h_{k+1}))$. For the above choice of basis functions and mesh ratio, we would choose $\rho_k = (\frac{1}{2})^p = \frac{1}{4}$, $p = 2$, which means that the residuals decrease in the continuation process $\Omega_{h_k} \to \Omega_{h_{k+1}}$ at the same ratio as the asymptotic discretization error in the "energy norm" $\|\cdot\|_V$, assuming that the exact solution $\hat{u} \in H^3(\Omega)$. After O((log h)) continuation steps, we have a residual error $O(h^p)$, $h \to 0$.

For an early paper on nested iterations, see [7]. For the use of a classical continuation process for the numerical solution of nonlinear boundary value problems, see [8]. Note that in that paper it has to be assumed that the discretization error for the nonlinear problem behaves like $O(h^p)$.

References

1. O. Axelsson, Conjugate gradient type methods for unsymmetric and inconsistent systems of linear equations, Linear Algebra and its Applications, 29 (1980), 1-16.

2. O. Axelsson, On multigrid methods of the two-level type. In Proceedings, Conference on multigrid methods, DFVLR, Köln-Porz, November 23-27th, 1981, Springer Verlag, to appear.

3. R.E. Bank and D.J. Rose, Global approximate Newton methods, Numer. Math. 37 (1981), 279-295.

4. R.S. Dembo, S.C. Eisenstat, and T. Steihaug, Inexact Newton Methods. Series # 47, School of Organization and Management, Yale University, 1980.

5. J.E. Dennis and J.J. Moré, A characterization of superlinear convergence and its application to quasi-Newton methods, Math. Comp. 28 (1974), 549-560.

6. L.V. Kantorovich, Functional analysis and applied mathematics, Uspekhi Mat. Nauk. 3 (1948), 89-185; English transl., Rep. 1509, National Bureau of Standards, Washington, D.C., 1952.

7. L. Kronsjö and G. Dahlquist, On the design of nested iterations for elliptic difference equations, BIT 11 (1971), 63-71.

8. L. Mansfield, On the solution of nonlinear finite element systems, SIAM J. Numer. Anal. 17 (1980), 752-765.

9. J.M. Ortega and W.C. Rheinboldt, Iterative solution of nonlinear equations in several variables, Academic Press, New York, 1970.

10. L.B. Rall, Computational solution of nonlinear operator equations, Wiley, New York, 1969.

MULTI-GRID SOLUTION OF CONTINUATION PROBLEMS

W. Hackbusch

Mathematisches Institut, Ruhr-Universität Bochum

Postfach 10 21 48, D-4630 Bochum 1, Germany

1. Introduction

The fast solution of *continuation problems*

(1.1) $L(u(\lambda),\lambda) = 0$

is discussed for the case that $L(\cdot,\lambda) = 0$ describes an elliptic boundary problem.

Multi-grid algorithms are known as efficient methods for solving linear as well as nonlinear elliptic equations $L(u) = 0$ [i.e. (1.1) for fixed λ]. Instead of concentrating to one discrete equation, the multi-grid iteration makes use of a sequence of discretizations

(1.2) $L_k(u_k) = 0 \qquad (0 \le k \le \ell)$

corresponding to grid sizes $h_o > h_1 > \ldots > h_{k-1} > h_k > \ldots > h_\ell$.

In § 2 we describe the *nonlinear multi-grid iteration* for a single equation $L(u) = 0$. Elsewhere (cf. [10, 13]) the convergence of the nonlinear multi-grid iteration is proved under conditions known from the linear multi-grid iteration. Here, we focus on the characteristic difficulties of nonlinear iterations. We try to describe a neighbourhood of the solution so that the multi-grid iteration is well-defined and that the iterates u_k^j converge to the solution u_k^*,

(1.3) $\|u_k^j - u_k^*\| \le \rho \, \|u_k^{j-1} - u_k^*\|$ with $\rho < 1$,

not leaving this neighbourhood.

The simplest application of an iteration for solving $L_\ell(u_\ell) = 0$ is to start with some guess u_ℓ^o and to apply a certain number of iterations resulting in u_ℓ^i. A more efficient approach is the *nested iteration* (cf. Kronsjö [16]):

1) $\tilde{u}_o$: approximate solution of $L_o(u_o) = 0$; $k := 0$;
2) $k := k+1$, define a starting iterate u_k^o by interpolation of $\tilde{u}_{k-1}$; compute $\tilde{u}_k = u_k^i$ by i steps of an iteration solving Eq. (1.2). If $k < \ell$ go to 2)

Starting at the coarsest grid, one approximates u_k by i iterations at all levels $k = 1,2,\ldots,\ell$. Thanks to the result computed at the previous level, one is able to start with a quite good guess u_k^o.

In § 2.4 we prove that any solver satisfying (1.3) combined with the nested iteration leads to approximations at level k with errors $O(h_k^\kappa \rho^i)$, where κ is the order of consistency and i is the number of iterations per level. More precisely, the error is bounded by the discretization error times a well-known factor $O(\rho^i)$. Obviously, the results of the nested iteration are superior to the naive application of i iterations (at level ℓ) yielding an error $O(\rho^i)$. It is to be noted that the nested iteration requires only little more operations than the naive iteration.

However, for the *continuation problem* (1.1) it is not optimal to apply the nested iteration described above for each value λ. In § 4 we propose a modified nested iteration that uses interpolated coarse-grid values at $\lambda = \lambda_{\nu+1}$ as well as results from the foregoing λ-value λ_ν.

This approach yields an error of $O(\Delta\lambda\, h_k^\kappa \rho^i)$ with $\Delta\lambda = \lambda_{\nu+1} - \lambda_\nu$, even though it is as cheap and as simple as the usual nested iteration. Thus, we gain an additional factor $\Delta\lambda$.

In § 4.3.5 we compare our nested iteration with the "frozen-τ technique" proposed by Brandt [5] and show that, in general, the former approach is superior.

Difficulties arising from *turning points* (limit points) of the branch are discussed in § 5. Introducing a new parameter, we sometimes have to look for a solution (u_ℓ, λ_ℓ) satisfying $L_\ell(u_\ell, \lambda_\ell) = 0$ and an additional scalar equation. For this extended problem we describe a modified multi-grid iteration.

Though only elliptic continuation problems are mentioned in this contribution, there are obvious extensions. If $L(\cdot,\lambda)$ is an *integral equation*, we can apply the (nonlinear) multi-grid iteration of the second kind for solving the discrete problems. This iteration has even better convergence results than (1.3) (cf. Hackbusch [11]). This multi-grid iteration of the second kind can also be applied to elliptic problems (cf. Hackbusch [9]).

Furthermore, the continuation problem $L(u,\lambda) = 0$ [solved at $\lambda = \lambda_o, \lambda_1, \ldots$] can be replaced by a *series of problems* $L^{(\nu)}(u) = 0$ $(\nu = 0,1,\ldots)$ having kindred solutions $u^{(\nu)}$. E.g., such problems arise for time-dependent (parabolic) problems, where elliptic problems have to be solved at each time step $t = \nu\Delta t$.

2. Case of a Single Nonlinear Problem

2.1 Continuous and Discrete Problems

Consider the problem (1.1) for a fixed λ. Let

$$(2.1) \qquad L(u) = 0$$

be a nonlinear boundary value problem (i.e., differential equation in Ω with boundary data on $\Gamma = \partial\Omega$.)The multi-grid approach requires the simultaneous use of different grid sizes

(2.2) $\quad h_0 > h_1 > \ldots > h_{k-1} > h_k > \ldots > h_\ell .$

The discretization of problem (2.1) corresponding to the step size h_k is denoted by

(2.3) $\quad L_k(u_k) = 0.$

This equation is also called the discretization 'at level k'. There is a double connection of the different levels. First, the multi-grid iteration for solving Eq. (2.3) at level $k^* \le \ell$ requires auxiliary equations

(2.4) $\quad L_k(u_k) = f_k$

with varying right-hand sides f_k for $k = 0,1,\ldots,k^*-1$ (cf. § 2.3). Second, we shall solve (2.3) for all $k = 0,1,\ldots,\ell$ as required by the nested iteration (cf. § 2.4).

2.2 Unique Solvability

It is not assumed that there is only one solution u^* of Eq. (2.1) and only one u_k^* satisfying (2.3). Other solutions may exist. Therefore, we have to fix a domain U_k and a range F_k of L_k such that

(2.5) $\quad L_k : U_k \to F_k$ is bijective

and $u_k^* \in U_k$ is the solution we are interested in.

In the following we have to curtail the domain. Let

$$U_k(r) := \{u_k : \|u_k - u_k^*\|_U \le r\}$$

be a sphere of radius r with respect to a norm $\|\cdot\|_U$ that will be used for the convergence estimate of the iteration, too. For example, $\|\cdot\|_U$ may be the Euclidean norm (ℓ_2-norm) or the discrete energy norm, etc.

Usually, we cannot expect that there is *one* $\varepsilon > 0$ with $U_k(\varepsilon) \subset U_k$ for all k. The reason is as follows. Assume that the coefficients of the differential equation (2.1) depend on u (or u_x, u_y). Then, reasonable neighbourhoods of u_k^* should be described by means of the supremum norm $\|\cdot\|_\infty$ (or supremum norm $\|\cdot\|_{1,\infty}$ of the values and first differences, resp.). But $\|\cdot\|_\infty$ (or $\|\cdot\|_{1,\infty}$) is not uniformly equivalent to, e.g., the Euclidean norm $\|\cdot\|_U$: $\|\cdot\|_\infty \le C\,\|\cdot\|_U$ does not hold with $C \ne C(k)$. But in the two-dimensional case ($\Omega \subset \mathbb{R}^2$) we have $\|\cdot\|_\infty \le Ch_k^{-1}\,\|\cdot\|_U$ (or $\|\cdot\|_{1,\infty} \le Ch_k^{-2}\,\|\cdot\|_U$) for the Euclidean norm $\|\cdot\|_U$.

Thus, there is a sequence of radii ε_k such that

(2.6) $\quad U_k(\varepsilon_k) \subset U_k.$

In the foregoing example the radii behave as $\varepsilon_k = \varepsilon h_k$ (or $\varepsilon_k = \varepsilon h_k^2$, resp.). Nevertheless, the case $\varepsilon_k = \varepsilon$ is possible. If for instance the iteration converges with respect to the discrete $H^2(\Omega)$ norm $\|\cdot\|_U$, the estimate $\|\cdot\|_\infty \le C\,\|\cdot\|_U$

(or $\|\cdot\|_{1,\infty} \leq C\,|\log h_k|\,\|\cdot\|_u$) implies $\varepsilon_k = \varepsilon$ (or $\varepsilon_k = \varepsilon\,/\,|\log h_k|$, resp.). The use of discrete H^2 norms requires some knowledge of the smoothness (regularity) of the discrete solution. For a discussion of regularity problems we refer to Hackbusch [12].

The image of $\mathcal{U}_k(r)$ is defined by

$$F_k(r) := \{f_k = L_k(u_k) : u_k \in \mathcal{U}_k(r) \cap \mathcal{U}_k\}.$$

We shall restrict L_k to $L_k : \mathcal{U}_k(\varepsilon_k) \to F_k(\varepsilon_k)$. ε_k is to be chosen so that (2.6) holds and that the iteration works.

2.3 Nonlinear Multi-Grid Iteration

2.3.1 Two-Grid Iteration

Let $k \in \{1,\ldots,\ell\}$ be fixed. The two-grid iteration for solving Eq. (2.4), $L_k(u_k) = f_k$, consists of smoothing parts and a coarse-grid correction. 'Smoothing' means the application of few iterations of a 'smoothing procedure', e.g., a nonlinear Gauß-Seidel iteration(cf. [13,18]). The coarse-grid correction in the two-grid case requires the exact solving of a nonlinear coarse-grid equation (2.4) at level k-1.

We denote the inverse of L_k by Φ_k:

$$\Phi_k(f_k) \text{ be solution of } L_k(u_k) = f_k.$$

Φ_k is well-defined on F_k with range $\mathcal{U}_k$ (cf. (2.5)). The solution u_k^* of the original equation (2.3) can be written as $\Phi_k(0)$.

The $(j+1)^{st}$ iterate u_k^{j+1} of the two-grid iteration is obtained from the jth iterate u_k^j by the following algorithm:

(2.7a) u_k' : result of smoothing applied to u_k^j ,

(2.7b) $\left.\begin{aligned} d_k &:= L_k(u_k') - f_k \quad \text{(defect of } u_k'\text{)}; \\ u_k'' &:= u_k' - \frac{1}{\sigma}\, p[\Phi_{k-1}(\tilde{f}_{k-1} + \sigma r d_k) - \Phi_{k-1}(\tilde{f}_{k-1})] \end{aligned}\right\}$ coarse-grid correction

(2.7c) u_k^{j+1} : result of smoothing applied to u_k'' .

p and r denote the prolongation (coarse-to-fine interpolation) and restriction (fine-to-coarse transfer). The coarse-grid correction depends on the values of $\sigma \in \mathbb{R}$ and $\tilde{f}_{k-1}$. $\tilde{f}_{k-1}$ is given by means of

(2.8) $\tilde{f}_{k-1} = L_{k-1}(\tilde{u}_{k-1})$,

where $\tilde{u}_{k-1}$ is to be chosen suitably. Therefore, the term $\Phi_{k-1}(\tilde{f}_{k-1})$ of (2.7b) requires no solving of a nonlinear equation. The coarse-grid correction can be rewritten as

(2.7b') $u_k'' := u_k' - \frac{1}{\sigma}\, p[\Phi_{k-1}(\tilde{f}_{k-1} + \sigma r d_k) - \tilde{u}_{k-1}]$.

In the *linear* case of $\mathcal{L}_k(u_k) \equiv L_k u_k - g_k$ the coarse-grid correction (2.7b) becomes

$$u_k'' = u_k' - pL_{k-1}^{-1}\, rd_k$$

and is independent of σ and $\tilde{f}_{k-1}$. Because of this fact, the nonlinear iteration (2.7) behaves asymptotically as the linear two-grid iteration.

Thus, we expect that

$$(2.9) \qquad \|u_k^{j+1} - \Phi_k(f_k)\|_U \leq \rho\, \|u_k^j - \Phi_k(f_k)\|_U, \ \rho < 1, \text{ for } u_k^j \in \mathcal{U}_k(\varepsilon_k),$$

provided that ε_k, d_k, σ and $\tilde{f}_{k-1}$ are sufficiently small. Indeed, the estimate (2.9) can be proved under conditions known from the linear case and under the assumption that $\mathcal{L}_k$ is differentiable at $u_k^* = \Phi_k(0)$ (cf. Hackbusch [10, 13]).

At present the only condition on $\tilde{f}_{k-1}$ is $\tilde{f}_{k-1} \in \mathcal{F}_{k-1}(\eta)$, $\eta < \varepsilon_{k-1}$. Then it is possible to choose σ with $\tilde{f}_{k-1} + \sigma rd_{k-1} \in \mathcal{F}_{k-1}(\varepsilon_{k-1})$ and the term $\Phi_{k-1}(\tilde{f}_{k-1} + \sigma rd_k)$ from (2.7b) is well-defined.

In order to repeat the iteration we have to ensure that the new iterate u_k^{j+1} belongs to $\mathcal{U}_k(\varepsilon_k)$, again.

Note 2.1 Assume one of the following two cases (2.10 a,b):

(2.10a) $f_k = 0$ and $u_k^o \in \mathcal{U}_k(\varepsilon_k)$,

(2.10b) $f_k \in \mathcal{F}_k(\varepsilon_k / 3)$ and $u_k^o \in \mathcal{U}_k(\varepsilon_k / 3)$,

where ε_k is chosen such that (2.6) and (2.9) hold. Then all iterates u_k^j remain in $\mathcal{U}_k(\varepsilon_k)$.

Proof. In case of (2.10a) the solution is $\Phi_k(f_k) = \Phi_k(0) = u_k^*$. By (2.9), the assumption $u_k^j \in \mathcal{U}_k(\varepsilon_k)$ implies $u_k^{j+1} \in \mathcal{U}_k(\varepsilon_k)$, directly. The second case (2.10b) is proved by induction. Assume $u_k^o,\ldots,u_k^j \in \mathcal{U}_k(\varepsilon_k)$. One concludes from (2.9) that

$$\begin{aligned}\|u_k^{j+1} - u_k^*\|_U &\leq \|u_k^{j+1} - \Phi_k(f_k)\|_U + \|\Phi_k(f_k) - u_k^*\|_U \leq \\ &\leq \rho^{j+1}\|u_k^o - \Phi_k(f_k)\|_U + \|\Phi_k(f_k) - u_k^*\|_U \leq \\ &\leq \rho^{j+1}(\|u_k^o - u_k^*\|_U + \|\Phi_k(f_k) - u_k^*\|_U) + \|\Phi_k(f_k) - u_k^*\|_U \leq \\ &\leq \rho^{j+1}(\varepsilon_k / 3 + \varepsilon_k / 3) + \varepsilon_k / 3 \leq \varepsilon_k;\end{aligned}$$

i.e., $u_k^{j+1} \in \mathcal{U}_k(\varepsilon_k)$, too. ∎

We recall that the sequence ε_k may behave as $\varepsilon_k = \varepsilon$, εh_k^α $(\alpha > 0)$, $\varepsilon/|\log h_k|$ etc. with ε small enough to ensure (2.9).

In the multi-grid case the solution $\Phi_{k-1}(\tilde{f}_{k-1} + \sigma rd_k)$ of the coarse-grid equation is approximated by the same method. Therefore, in view of (2.10b) one has to fulfil $\tilde{f}_{k-1} + \sigma rd_k \in \mathcal{F}_{k-1}(\varepsilon_{k-1} / 3)$. This requirement leads to the following choice of $\tilde{f}_{k-1}$ (or $\tilde{u}_{k-1} = \Phi_{k-1}(\tilde{f}_{k-1})$, resp.) and σ. Assume that (2.11),

$$(2.11) \qquad \|\Phi_k'(f_k)d_k\|_U \leq C_o\|d_k\|_F \text{ for all } f_k \in \mathcal{F}_k(\varepsilon_k), \text{ all } d_k, \text{ and } 1 \leq k \leq \ell,$$

is valid for a (computable) norm $\|\cdot\|_F$ (e.g., Euclidean norm). Note that the derivative $\Phi_k' = \partial\Phi_k / \partial f_k$ becomes L_k^{-1} in the linear case of $L_k(u_k) \equiv L_k u_k - g_k$.

Note 2.2 Choose $\tilde{u}_{k-1} = \Phi_{k-1}(\tilde{f}_{k-1})$ and σ according to

(2.12) $\tilde{u}_{k-1} \in U_{k-1}(\varepsilon_{k-1} / 6)$, $|\sigma| \leq \sigma_{k-1} / \|rd_k\|_F$ with $\sigma_{k-1} = \frac{\varepsilon_{k-1}}{6C_o}$.

Then the right-hand side $f_{k-1} := \tilde{f}_{k-1} + \sigma rd_k$ belongs to $F_{k-1}(\varepsilon_{k-1} / 3)$.

The case of $rd_k = 0$ can be neglected since then the coarse-grid correction (2.7b) can be omitted.

2.3.2 Nonlinear Multi-Grid Iteration

Approximating $\Phi_{k-1}(\tilde{f}_{k-1} + \sigma rd_k)$ from (2.7b) by γ iterations of the same iteration at the coarser grid, we obtain the multi-grid iteration. It can be defined recursively as follows. For the coarsest grid (k = 0) define some suitable iteration

(2.13a) $u_o^{j+1} = \phi_o(u_o^j, f_o)$

that converges to $\Phi_o(f_o)$ (at least for $f_o \in F_o(\varepsilon_o)$). Having defined the iteration ϕ_{k-1} at level k-1, we describe the iteration ϕ_k at level k by

(2.13ba) u_k' : result of smoothing applied to u_k^j;

(2.13bb) $d_k := L_k(u_k') - f_k$; $\sigma := \sigma_{k-1} / \|rd_k\|_F$;

(2.13bc) $v_{k-1}^o := \tilde{u}_{k-1}$ $\quad$ ($\tilde{u}_{k-1}$ from (2.8));

(2.13bd) $v_{k-1}^{\mu} = \phi_{k-1}(v_{k-1}^{\mu-1}, \tilde{f}_{k-1} + \sigma rd_k)$ $\quad$ for $\mu := 1,\ldots,\gamma$;

(2.13be) $u_k'' := u_k' - p(v_{k-1}^{\gamma} - \tilde{u}_{k-1}) / \sigma$;

(2.13bf) $\phi_k(u_k^j, f_k) := u_k^{j+1}$: result of smoothing applied to u_k''.

The number σ_{k-1} involved in (2.13bb) is defined in (2.12). The function $\tilde{u}_{k-1}$ must be given in before. Therefore, $\tilde{f}_{k-1} = L_{k-1}(\tilde{u}_{k-1})$ has to be evaluated only once. The number γ is 1 or 2.

By usual arguments (cf. [13]) the convergence (2.9) of the two-grid iteration leads to convergence of the multi-grid iteration:

(2.14) $\|\phi_k(u_k,f_k) - \Phi_k(f_k)\|_U \leq \rho'\|u_k - \Phi_k(f_k)\|_U$, $\rho' < 1$

for $(u_k,f_k) \in U_k(\varepsilon_k) \times F_k(\varepsilon_k)$ and auxiliary values $\tilde{u}_o,\ldots,\tilde{u}_{k-1}$ fulfilling (2.12). Note 2.2 ensures that all iterates remain in the respective subsets. We summarize:

Note 2.3 Assume that

- the nonlinear iteration ϕ_o at the coarsest grid have a contraction number $\rho' < 1$,
- $\tilde{u}_m (0 \leq m \leq k-1)$ satisfy (2.12),
- convergence estimate (2.9) holds for the two-grid iteration with ρ small enough,
- $\gamma \geq 2$.

Then the multi-grid iteration (2.13) converges; more precisely: (2.14) is valid for all $u_k \in \mathcal{U}_k(\varepsilon_k)$, $f_k \in \mathcal{F}_k(\varepsilon_k)$. If the starting guess u_k^o and the right-hand side f_k satisfy (2.10a) or (2.10b), then all iterates $u_k^j = \phi_k(u_k^{j-1}, f_k)$ remain in $\mathcal{U}_k(\varepsilon_k)$.

2.3.3 Other choice of σ and $\tilde{u}_{k-1}$

Note 2.3 shows the *local* convergence of the iteration (2.13). The initial value u_k^o must not differ from u_k^* by more than ε_k. Even in cases, where the problems (2.4), $\mathcal{L}_k(u_k) = f_k$, are uniquely solvable for all f_k, the radius ε_k might be small. Are there versions guaranteeing global convergence?

Consider again the two-grid iteration (2.7) and set

(2.15) $\quad \sigma = -1, \; \tilde{u}_{k-1} = \hat{r}u_k' = \hat{r}\Phi_k(f_k + d_k),$

where $\hat{r}$ is some restriction (possibly $\hat{r} \neq r$). $\tilde{u}_{k-1}$ is cheaply attainable; however, $\tilde{f}_{k-1} = \mathcal{L}_{k-1}(\tilde{u}_{k-1})$ has to be renewed in the next iteration since u_k' changes. The choice (2.15) leads to Brandt's FAS version (cf. [4]).

The result u_k'' of (2.7b) differs from the exact solution $\Phi_k(f_k)$ by

$$v_k = u_k' - \Phi_k(f_k) + p[\Phi_{k-1}(\tilde{f}_{k-1} - rd_k) - \Phi_{k-1}(\tilde{f}_{k-1})] =$$
$$= \Phi_k(f_k + d_k) - \Phi_k(f_k) - p[\Phi_{k-1}(\tilde{f}_{k-1}) - \Phi_{k-1}(\tilde{f}_{k-1} - rd_k)] =$$
$$= [\int_o^1 \{\Phi_k'(f_k + sd_k) - p\Phi_{k-1}'(\tilde{f}_{k-1} - rd_k + srd_k)r\}ds]d_k.$$

If one is able to prove an 'approximation property' (cf. [10,13]) for $\Phi_k'(f_k + sd_k) - p\Phi_{k-1}'(\tilde{f}_{k-1} - rd_k + sd_k)r$ uniformly for all $s \in [0,1]$, $f_k + sd_k \in F_k(\varepsilon_k)$ with *large* ε_k, one could prove global convergence of the two-grid iteration (2.7) and the multi-grid iteration (2.13) with σ and $\tilde{u}_{k-1}$ from (2.15).

2.3.4 Newton-Multigrid Iteration

Linear multi-grid algorithms can be used in combination with Newton's iteration. The prototype of the algorithm is (2.16):

(2.16a) compute $d_k := \mathcal{L}_k(u_k^j)$; set $L_k \approx \mathcal{L}_k'(u_k^j)$;

(2.16b) apply one step of the (linear) multi-iteration to $L_k v_k = d_k$ starting with $v_k^o = 0$ resulting in v_k^1;

(2.16c) $u_k^{j+1} := u_k^j - v_k^1$

(cf. Bank [1,2]). Under natural conditions one obtains convergence:

(2.17) $\quad \|u_k^{j+1} - u_k^*\|_U \le \rho \, \|u_k^j - u_k^*\|_U, \; \rho < 1$, for all $u_k^j \in \mathcal{U}_k(\varepsilon_k)$.

Usually, $\mathcal{U}_k(\varepsilon_k)$ is a small neighbourhood, since otherwise the Newton iteration is not convergent (to u_k^*) or too slow.

2.3.5 Nonlinear Iterative Solver

In the sequel we shall make no use of the special nature of the nonlinear multi-grid iteration (2.13) or of the Newton variant (2.16). Further, it is of no more interest that we are able to solve the perturbed equation $L_k(u_k) = f_k \neq 0$, too. The iterations (2.13) as well as (2.16) are iterative solvers of $L_k(u_k) = 0$. We denote the iteration by

(2.18) $\quad u_k^{j+1} = \phi_k(u_k^j)$.

The desired properties of ϕ_k are

(2.19a) $\quad \phi_k$ is defined on $U_k(\varepsilon_k)$,

(2.19b) $\quad \|\phi_k(u_k) - u_k^*\|_U \leq \rho \|u_k - u_k^*\|_U$ with $\rho < 1$ for all $u_k \in U_k(\varepsilon_k)$,

where u_k^* is the only solution of $L_k(u_k)=0$ in $U_k(\varepsilon_k)$. The assumption 'ρ being independent of k' is typical for multi-grid iterations. Therefore, we shall not bother about contraction number dependent on k, although the algorithms described in [9,11] have rates $\rho_k = \rho h_k^\kappa \to 0$ $(h_k \to 0)$.

2.4 Nested Iteration for Single Equations

Any iterative process can be improved by preparing good starting values. Let $\tilde{p}$ be a prolongation from level k-1 to k (coarse-to-fine interpolation). $\tilde{p}$ may coincide with p from the multi-grid algorithm or it may be more accurate. The algorithm reads as follows:

(2.20a) $\quad \tilde{u}_o$:= approximate solution of $L_o(u_o) = 0$;

(2.20b) $\quad$ <u>for</u> k := 1 (1) ℓ <u>do</u>

(2.20c) $\quad$ <u>begin</u> [$\tilde{f}_{k-1} := L_{k-1}(\tilde{u}_{k-1})$;]

(2.20d) $\quad\quad \tilde{u}_k := \tilde{p}\tilde{u}_{k-1}$;

(2.20c) $\quad\quad$ <u>for</u> j := 1(1) i <u>do</u> $\tilde{u}_k := \phi_k(\tilde{u}_k)$

<u>end</u>;

The statement (2.20c) is put into brackets, since $\tilde{f}_{k-1}$ is needed for the multi-grid version (2.13) but is of no purpose for other solvers ϕ_k. If ϕ_k is a multi-grid iteration, the algorithm (2.20) is also called 'full multi-grid method' (cf.[4,18]).

To prepare the next theorem we have to define the *relative discretization error*

(2.21) $\quad \|\tilde{p}u_{k-1}^* - u_k^*\|_U \leq C_1 h_k^\kappa \qquad \left\{ \begin{array}{l} \kappa : \text{consistency order} \\ u_k^*: \text{solution of (2.3)} \end{array} \right.$

and the constants C_{2o}, C_{21}, C_2:

(2.22a) $\quad \|\tilde{p}v_{k-1}\|_U \leq C_{2o}\|v_{k-1}\|_U$ for all v_{k-1} ,

(2.22b) $\quad h_{k-1} \leq C_{21}h_k$,

(2.22c) $C_2 := C_{2o} C_{21}^{\kappa}$.

Note 2.4 The constant C_2 is available. Most $\tilde{p}$ satisfy (2.22a) with $C_{2o} = 1$. The usual ratio h_{k-1} / h_k is $C_{21} = 2$, leading to $C_2 = 2^{\kappa}$.

The starting guess $\tilde{\tilde{p}}u_{k-1}$ must belong to $\mathcal{U}_k(\varepsilon_k)$. Otherwise, ϕ_k may fail to converge to u_k^*. Setting $\tilde{u}_{k-1} = u_{k-1}^*$ we are led to

(2.23) $C_1 h_k^{\kappa} < \varepsilon_k \qquad (1 \leq k \leq \ell)$.

We recall that ε_k may depend on h_k as $\varepsilon_k = \varepsilon h_k^{\alpha}$. For $\alpha \leq \kappa$ inequality (2.23) becomes $C_1 h_1^{\kappa-\alpha} < \varepsilon$, otherwise $C_1 h_o^{\kappa-\alpha} < \varepsilon$.

Theorem 2.5 Let the nonlinear iteration ϕ_k satisfy the convergence condition (2.19) and assume (2.21), (2.22), (2.23). By (2.23) the number i (cf. (2.20c)) can be chosen so that

(2.24) $C_2 \rho^i \leq 1 - C_1 h_k^{\kappa} / \varepsilon_k \qquad (1 \leq k \leq \ell)$.

Then, the nested iteration (2.20) is well-defined and produces results $\tilde{u}_k$ with

(2.25) $\|\tilde{u}_k - u_k^*\|_U \leq C_3(\rho)\, C_1 h_k^{\kappa}$ with $C_3(\rho) = \dfrac{\rho^i}{1 - C_2 \rho^i}$ for $0 \leq k \leq \ell$,

provided that the starting value $\tilde{u}_o$ at the coarsest grid satisfies (2.25).

Proof. By assumption, (2.26) holds for $k = 0$. Let (2.26) be valid for $k-1 < \ell$. The starting value at level k is $u_k^o = \tilde{\tilde{p}}u_{k-1}$. The estimate

$$\|u_k^o - u_k^*\|_U \leq \|\tilde{p}u_{k-1}^* - u_k^*\|_U + \|\tilde{p}(\tilde{u}_{k-1} - u_{k-1}^*)\|_U \leq$$
$$\leq C_1 h_k^{\kappa} + C_{2o}\|\tilde{u}_{k-1} - u_{k-1}^*\|_U \leq C_1 h_k^{\kappa} + C_{2o} C_3(\rho)\, C_1 h_{k-1}^{\kappa} \leq$$
$$\leq [1 + C_{2o} C_{21}^{\kappa} C_3(\rho)]\, C_1 h_k^{\kappa} = [1 + C_2 C_3(\rho)]\, C_1 h_k^{\kappa} = C_1 h_k^{\kappa} / [1 - C_2 \rho^i]$$

together with (2.24) ensures $u_k^o \in \mathcal{U}_k(\varepsilon_k)$. Thus, (2.19) implies

$$\|\tilde{u}_k - u_k^*\|_U \leq \rho^i \|u_k^o - u_k^*\|_U \leq C_3(\rho) C_1 h_k^{\kappa}$$

proving (2.25) for level k, too. ∎

Corollary 2.6 The algorithm (2.13) at level k needs auxiliary values $\tilde{u}_m \in \mathcal{U}_m(\varepsilon_m/6)$ [$0 \leq m \leq k-1$; compare (2.12) and Note 2.3]. $\tilde{u}_k \in \mathcal{U}_k(\varepsilon_k/6)$ holds for the results of the nested iteration (2.20) if in addition

(2.26) $\rho^i \leq 1/6$.

A careful analysis reveals that (2.12) can be replaced by $\tilde{u}_k \in \mathcal{U}_k(\theta\varepsilon_k)$, $\theta = (5-\rho) / [6(2\rho + 1)]$, with same σ_{k-1}. Weakening (2.26) one obtains that (2.26') implies $\tilde{u}_k \in \mathcal{U}_k(\theta\varepsilon_k)$:

(2.26') $\dfrac{6\rho^i}{1 - C_2 \rho^i} \dfrac{2\rho+1}{5-\rho} C_1 h_k^{\kappa} \leq \varepsilon_k$.

In most of the cases it is reasonable to require that the approximation $\tilde{u}_k$ is not too good. Since u_k^* and u^* differ by the discretization error $O(h_k^\kappa)$, one should terminate the iteration when the iteration error $\tilde{u}_k - u_k^*$ is of size $O(h_k^\kappa)$. Theorem 2.5 shows that just this error estimate is guaranteed by the nested iteration. Further, the relation between the relative discretization error (bound $C_1 h_k^\kappa$) and the iteration error can be controlled. Its ratio is $C_3(\rho)$ depending on C_2 and ρ, only.

Note 2.7 $C_3(\rho)$ can be determined during the computation. For C_2 compare Note 2.4. Estimations of ρ can be obtained by observing the multi-grid convergence.

The naive approach would be to take some starting value u_ℓ^o (with the problem to ensure $u_\ell^o \in \mathcal{U}_\ell(\varepsilon_\ell)$) and to iterate the process ϕ_ℓ several times. But then $i = O(|\log h_k|)$ iterations are required instead of a small fixed number i in case of the nested iteration. Moreover, the control of the iteration error by $\rho^i \| u_k^o - u_k^* \|_U$ is difficult, since the absolute size of the discretization error $C_1 h_k^\kappa$ is usually unknown. We summarize: The nested iteration has two advantages. It is extremely cheap and it yields an automatic control of the iteration error.

We add the comment, that the nested iteration gives reasonable results even if $C_2\rho^i > 1$ [i.e. (2.24) violated], provided that $\tilde{p}u_{k-1} \in \mathcal{U}_k(\varepsilon_k)$ $(1 \le k \le \ell)$ remains valid.

The difficulty of the nested iteration in the nonlinear case is the requirement (2.23): $C_1 h_k^\kappa < \varepsilon_k$. Assuming $\varepsilon_k = \varepsilon h_k^\alpha$ $(\alpha \le \kappa)$ we see that the most restrictive inequality is (2.23) at level $k = 1$: $C_1 h_1^\kappa < \varepsilon_1$. Hence, a misfortune is expected - if at all - at the lowest levels. If $\tilde{p}u_o^*$ is not contained in the region of contraction of ϕ_1, the nested iteration must break down. Then, it may happen that the iteration ϕ_1 diverges or that the iterates go astray towards another (undesired) solution u_1^{**} of $L_1(u_1) = 0$. One remedy would be to omit h_o and to take h_1 as new coarsest grid size h_o. But then the computational complexity increases and it is still a problem to find a suitable starting guess u_1^o. Another remedy is to consider the homotopy $L_1(u_1(\lambda),\lambda) \equiv L_1(u_1) - (1-\lambda)f_1 = 0$ with $f_1 = L_1(\tilde{p}u_o)$. Then $u_1(0) = \tilde{p}u_o$ is known and one may hope that the branch connects $\tilde{p}u_o$ and $u_1(1) = u_1^*$. This last comment leads us to continuation methods.

It will turn out that difficulties as discussed above do not exist for a modified nested iteration applied to continuation problems.

3. Nonlinear Problem Depending on a Parameter

3.1 Continuous Problem

Often, a boundary value problem depends on a parameter λ:

(3.1) $$L(u(\lambda),\lambda) = 0,$$

where λ varies in some interval I. An example is the Dirichlet problem

(3.2) $$-\Delta u(\lambda) + g(u(\lambda),\lambda) = 0 \text{ in } \Omega,\ u = 0 \text{ on } \Gamma = \partial\Omega.$$

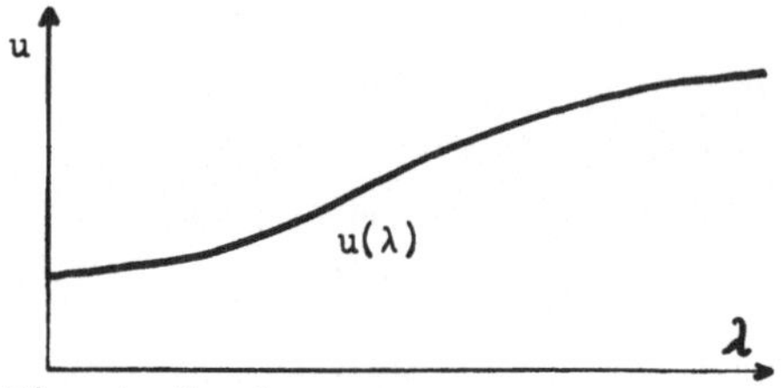

Fig. 1 Regular case

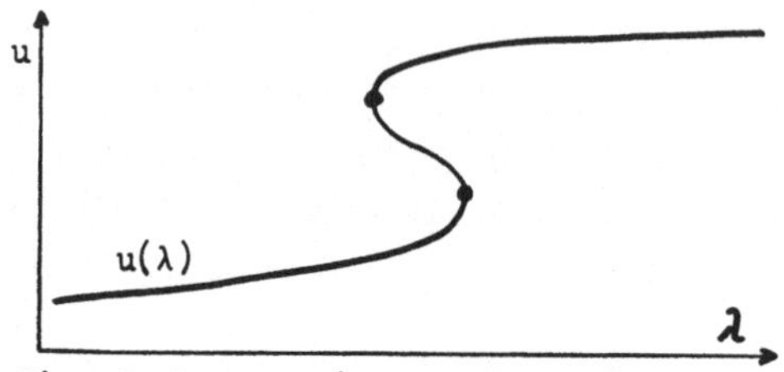

Fig. 2 Branch with turning points

In § 4 we shall concentrate on the regular case (cf. Fig. 1), where the desired branch $u(\lambda)$ is a function of λ. In this case λ is a reasonable parameter for describing the branch. If turning (limit) points occur (cf. Fig. 2), well-known techniques can be used as mentioned in § 5.

We assume that one is interested in following one branch $u(\lambda)$ and obtaining numerical results for several values $\{\lambda_o, \lambda_1, \ldots\} \subset I$.

3.2 Discrete Problems

Similarly as in § 2, the discrete problem at level k (corresponding to step size h_k) is denoted by

$$(3.3) \qquad \mathcal{L}_k(u_k(\lambda), \lambda) = 0.$$

Again, we shall first consider the regular case as depicted in Fig. 3. It will be required that not only $u_{k-1}(\lambda)$ approximates $u_k(\lambda)$ but also $\frac{\partial}{\partial\lambda} u_{k-1}(\lambda)$ is close to $\frac{\partial}{\partial\lambda} u_k(\lambda)$. We note that $v_k = \frac{\partial}{\partial\lambda} u_k(\lambda)$ is the solution of the linear problem

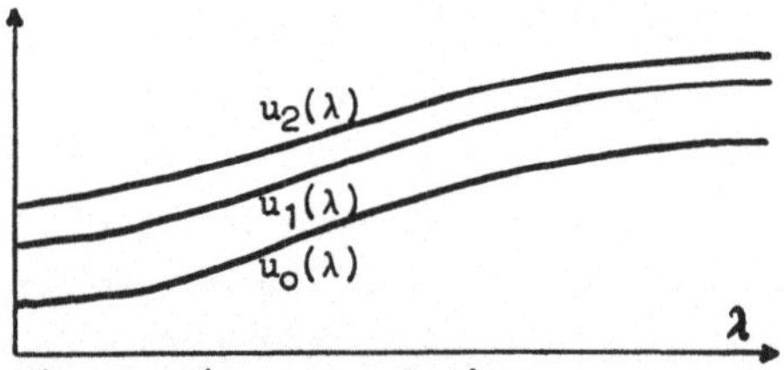

Fig. 3 Discrete solutions

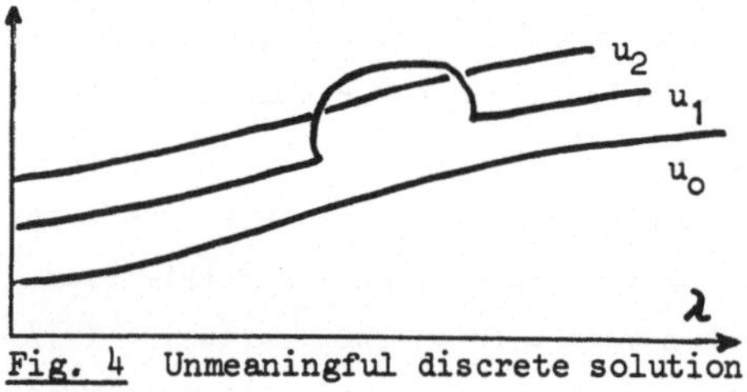

Fig. 4 Unmeaningful discrete solution

$$(3.4) \qquad L_k(u_k(\lambda), \lambda)\, v_k = g_k(u_k(\lambda), \lambda) \quad \left(L_k = \frac{\partial \mathcal{L}_k}{\partial u_k}\ , \ g_k = -\frac{\partial \mathcal{L}_k}{\partial \lambda}\right)$$

and thereby $v_k \approx v_{k-1}$ may be expected. However, it is known that unreasonabel discrete solutions might occur or that a well-behaved discrete solution $u_k(\lambda)$ can partly (for $\lambda \in I' \subset I$ and one level,only)become unmeaningful(cf. Fig. 4). Examples are described, e.g., by Bigge and Bohl [3] .

In the latter case the proposed nested iteration of § 4 will have difficulties. But this fact can be regarded as an advantage: Only by the simultaneous use of several levels wrong discrete branches can be detected.

3.3 Unique Solvability

It is not assumed that there is only one solution $u(\lambda)$. Fig. 1 shows the branch we want to follow; other branches may exist. Therefore, we have to fix a domain $\mathcal{U}_k^\lambda$ containing $u_k(\lambda)$ and a range F_k^λ of $L_k(.,\lambda)$ such that

$$L_k(.,\lambda): \mathcal{U}_k^\lambda \to F_k^\lambda \quad \text{be bijective for } \lambda \in I$$

(cf. (2.5)). We must ensure that during the (approximate) computation of $u_k(\lambda)$ we do not leave $\mathcal{U}_k^\lambda$. This is an important restriction when diam $\mathcal{U}_k^\lambda$ is smaller than the desired accuracy.

Again, spheres

$$\text{(3.5a)} \qquad \mathcal{U}_k^\lambda(r) = \{u_k : \|u_k - u_k(\lambda)\|_U \leq r\}$$

and the images

$$\text{(3.5b)} \qquad F_k^\lambda(r) = \{f_k = L_k(u_k,\lambda) : u_k \in \mathcal{U}_k^\lambda(r) \cap \mathcal{U}_k^\lambda\}$$

are introduced. As in § 2.2 we consider neighbourhoods $\mathcal{U}_k^\lambda(\varepsilon_k)$. For simplicity, ε_k is assumed to be independent of λ. By the same reason all estimates of discretization errors etc. are discribed by constants C independent of λ. Of course, all these numbers can depend on λ, and it is not difficult to extend the following results to this case.

3.4 Nonlinear Solver

As in § 2 our analysis is not only true for a special version of the multi-grid iteration, but for all iterative solvers that converge with uniform contraction number in $\mathcal{U}_k^\lambda(\varepsilon_k)$.

The iteration is denoted by ϕ_k:

$$\text{(3.6)} \qquad u_k^{j+1} = \phi_k(u_k^j,\lambda) \qquad (\lambda \in I).$$

We require

$$\text{(3.7a)} \qquad \phi_k \text{ be defined for } u_k \in \mathcal{U}_k^\lambda(\varepsilon_k), \qquad \lambda \in I,$$

$$\text{(3.7b)} \qquad \|\phi_k(u_k,\lambda) - u_k(\lambda)\|_U \leq \rho\, \|u_k - u_k(\lambda)\|_U \text{ with } \rho < 1$$
$$\text{for all } k \in \{1,\dots,\ell\},\ u_k \in \mathcal{U}_k^\lambda(\varepsilon_k),\ \lambda \in I.$$

It is well-known for multi-grid iterations that ρ is independent of k. However, ρ may depend on λ. As mentioned above, we use a uniform rate ρ only for the sake of simplicity.

4. Nested Iteration for Continuation Problems

4.1 Discussion of Various Approaches

Let us assume that we enter the problem at some $\lambda_o \in I$, where all discrete solutions $u_k(\lambda_o)$ are computed. We can proceed in two ways. Either we follow all branches $u_k(\lambda)$ $(o \leq k \leq \ell)$ or we compute $u_\ell(\lambda)$ at the finest level, only. The first approach yields a better control of the discretization errors and the reliability of the results. Hence, let us assume that we compute $\{u_k(\lambda): o \le k \le \ell\}$ for $\lambda = \lambda_o, \lambda_1, \dots, \lambda_\nu$. The basic problem is to determine $\{u_k(\lambda_{\nu+1}): o \le k \le \ell\}$ knowing $u_k(\lambda_\nu)$ (cf. Fig. 5).

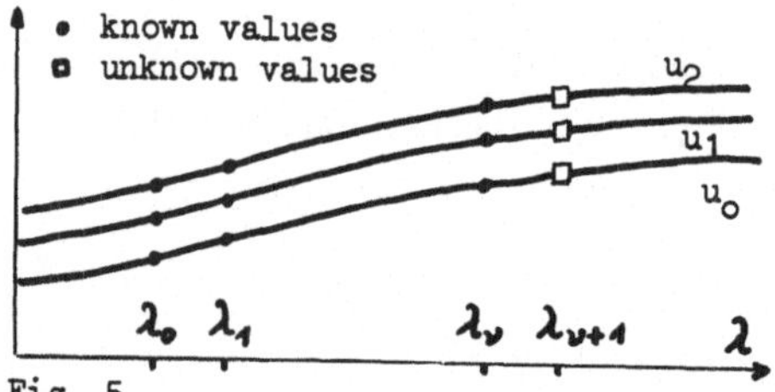

Fig. 5

The following requirements are imposed upon the algorithm:

(i) all discrete solutions $u_k(\lambda_{\nu+1})$, $o \le k \le \ell$, are to be computed;

(ii) the algorithm must be cheap;

(iii) as for the nested iteration of § 2 the iteration error $\tilde{u}_k(\lambda_{\nu+1}) - u_k(\lambda_{\nu+1})$ should be proportional to the discretization error;

(iv) the algorithm must be well-defined (at least for $\lambda_{\nu+1} - \lambda_\nu$ sufficiently small); it must be ensured that we do not leave the branch $u_k(\lambda)$.

Requirement (i) is discussed above. The other claims are obvious.

The naive application of the nested iteration (2.20) of § 2.4 to the problem $L_k(u_k) := L_k(u_k, \lambda_{\nu+1}) = 0$ would satisfy (i), (ii) and (iii). But (iv) is not fulfilled since this approach breaks down when (2.23) is violated, as explained at the end of § 2.4. The nested iteration (2.20) disregards the known values at λ_ν.

The given value $u_k(\lambda_\nu)$ can be used as starting value of the iteration ϕ_k at $\lambda_{\nu+1}$. This technique and improved versions are well-known for the one-grid case (cf. Deuflhard [8]). The latter approach has the benefit of property (iv). If $\lambda_{\nu+1} - \lambda_\nu$ is sufficiently small, $u_k(\lambda_\nu)$ belongs to $\mathcal{U}_k^{\lambda_{\nu+1}}(\varepsilon_k)$. To satisfy claim (i), one has to repeat the continuation for all levels $k \in \{0, \dots, \ell\}$. Hence, it is doubtful whether the algorithm is as cheap as possible (cf. (ii)). Finally, the initial error $u_k^o - u_k(\lambda_{\nu+1}) \approx u_k(\lambda_\nu) - u_k(\lambda_{\nu+1})$ is of order $O(\lambda_{\nu+1} - \lambda_\nu)$. A fixed number of iterations does not yield an error proportional to the discretization error $O(h_k^\kappa)$ as requested by (iii).

The first method proceeds at $\lambda = \lambda_{\nu+1}$ upwards from the coarsest grid to the finer ones, while the second approach moves from the left ($\lambda = \lambda_\nu$) to the right ($\lambda = \lambda_{\nu+1}$). The nested iteration of the subsequent section combines both aspects and inherits the advantages of both methods. Another approach, the 'frozen-τ technique' of Brandt [5], is discussed in § 4.3.5.

4.2 Continuation by Nested Iteration

Set $\lambda' := \lambda_\nu$ and $\lambda := \lambda_{\nu+1}$ and assume that approximations $\tilde{u}'_k$ of $u_k(\lambda')$ $(0 \le k \le \ell)$ are known (cf. Fig. 6).

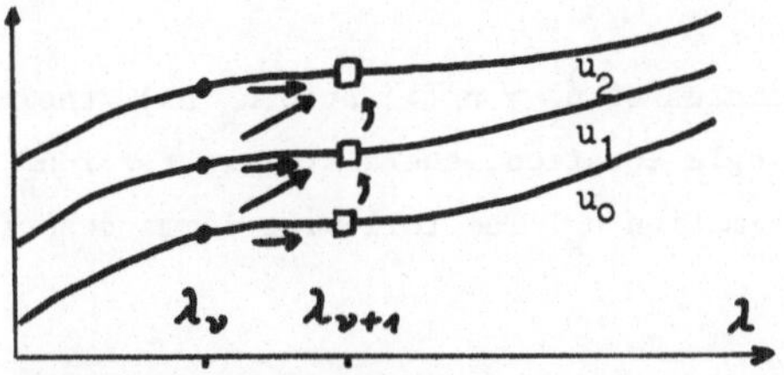

Fig. 6 Solution process

Approximations $\tilde{u}_k$ of $u_k(\lambda)$ at $\lambda = \lambda_{\nu+1}$ are computed by the nested iteration (4.1):

(4.1a) compute $\tilde{u}_o$ by some continuation method,

<u>for</u> $k := 1(1)\ \ell$ <u>do</u>

<u>begin</u>

(4.1b) $\tilde{u}_k := \tilde{u}'_k + \tilde{p}(\tilde{u}_{k-1} - \tilde{u}'_{k-1})$;

(4.1c) <u>for</u> $j := 1(1)\ i$ <u>do</u> $\tilde{u}_k := \phi_k(\tilde{u}_k, \lambda)$

$$\left(\begin{array}{l}\tilde{u}_k = \tilde{u}_k(\lambda)\\ \tilde{u}'_k = \tilde{u}_k(\lambda')\end{array}\right)$$

<u>end;</u>

As the nonlinear multi-grid iteration requires some other iteration ϕ_o at the lowest level, we need another continuation process in (4.1a). This continuation can be the iteration $\phi_o(\cdot,\lambda)$ with starting value u'_o or it may use intermediate solutions $u_o(\hat{\lambda})$, $\lambda' < \hat{\lambda} < \lambda$. It is plain that continuation at level $k = 0$ is the cheapest one (cf. claim (ii)). A possible counter-argument is that the solution $u_o(\lambda)$ at the coarsest grid might behave not as the finer-grid solutions and thereby it should not be used for computing u_1. However, if this happens one ought to cancel the lowest level, since one must fear that the level $k = 0$ causes difficulties also in the usual multi-grid iteration ϕ_k.

The construction (4.1b) for the starting guess $\tilde{u}_k$ is suggested by $u_k(\lambda) - u_k(\lambda') \approx \tilde{p}[u_{k-1}(\lambda) - u_{k-1}(\lambda')]$. Indeed, the difference is

$$\delta_k(\lambda,\lambda') := u_k(\lambda) - u_k(\lambda') - \tilde{p}[u_{k-1}(\lambda) - u_{k-1}(\lambda')] =$$

(4.2)

$$= \int_\lambda^{\lambda'} [v_k(t) - \tilde{p}\, v_{k-1}(t)]\,dt \text{ with } v_k(\lambda) := \frac{\partial}{\partial k} u_k(\lambda).$$

Assuming $v_k(\lambda)$ to be smooth (cf. § 3.2 and (3.4)) one obtains

$$v_k - \tilde{p}\, v_{k-1} = O(h_k^\kappa)$$

as in the case of the usual nested iteration (2.20), where $u_k - \tilde{p}\, u_{k-1} = O(h_k^\kappa)$. Thus, (4.2) shows

$$\delta_k(\lambda,\lambda') = O(h_k^\kappa \cdot \Delta\lambda) \text{ with } \Delta\lambda = \lambda-\lambda' \ (= \lambda_{\nu+1} - \lambda_\nu),$$

i.e.

$$(4.3)\qquad \|u_k(\lambda) - u_k(\lambda') - \tilde{p}[u_{k-1}(\lambda) - u_{k-1}(\lambda')]\|_U \le C_1\, h_k^\kappa\, \Delta\lambda.$$

Inequality (4.3) replaces the estimate $\|u_k - \tilde{p}u_{k-1}\|_U \le C_1\, h_k^\kappa$ in case of the usual nested iteration (2.20).

Can we obtain error estimates of $\tilde{u}_k - u_k(\lambda)$ by $O(h_k^\kappa\, \Delta\lambda)$, too? In contrast to the nested iteration for a single equation, there are errors of $\tilde{u}_k'$ at λ' shifting into errors of the new approximation $\tilde{u}_k$. The following lemma describes how these errors are diminished.

Lemma 4.1 Suppose

- (3.7a,b) [$\phi_k(.,\lambda)$ have contraction number ρ],
- (2.22a-c) [estimates of $\tilde{p}$ and h_{k-1}/h_k by C_{2o} and C_{21}; $C_2 = C_{2o}C_{21}^\kappa$],
- (4.3) [discretization error of $\partial u_k/\partial\lambda$].

Choose the iteration number i [cf. (4.1c)] such that

$$(4.4a)\qquad C_2\rho^i < 1 \qquad (C_2 \text{ from (2.22c)},\ \rho \text{ from (3.7b)})$$

and assume

$$(4.4b)\qquad [C_1\, \Delta\lambda + (1 + C_2)\eta']\, h_k^\kappa \,/\, [1 - C_2\rho^i] \le \varepsilon_k \qquad (1 \le k \le \ell)$$

with $\Delta\lambda = \lambda-\lambda'$ and η' involved in (4.5a). Assume that the errors of $\tilde{u}_k'$ satisfy

$$(4.5a)\qquad \|\tilde{u}_k' - u_k(\lambda')\|_U \le \eta' h_k^\kappa \qquad (0 \le k \le \ell)$$

and that $\tilde{u}_o$ from (4.1a) fulfils

$$(4.5b)\qquad \|\tilde{u}_o - u_o(\lambda)\|_U \le \eta h_o^\kappa$$

with

$$(4.5c)\qquad \eta = \frac{(1+C_2)\eta' + C_1\Delta\lambda}{1-C_2\rho^i}\,\rho^i.$$

Then the nested iteration (4.1) is well-defined and produces approximations $\tilde{u}_k$ with

$$(4.6)\qquad \|\tilde{u}_k - u_k(\lambda)\|_U \le \eta h_k^\kappa \qquad (0 \le k \le \ell).$$

Proof. (4.5b) implies (4.6) for k=0. Assume (4.6) for k-1. The error of the starting value $u_k^o := \tilde{u}_k' + \tilde{p}(\tilde{u}_{k-1} - \tilde{u}_{k-1}')$ from (4.1b) is

$$\|u_k^o - u_k(\lambda)\|_U = \|\{u_k(\lambda') - u_k(\lambda) - \tilde{p}[u_{k-1}(\lambda') - u_{k-1}(\lambda)]\} + \{\tilde{u}_k' - u_k(\lambda')\} -$$
$$- \tilde{p}[\{\tilde{u}_{k-1}' - u_{k-1}(\lambda')\} - \{\tilde{u}_{k-1} - u_{k-1}(\lambda)\}]\|_U \le$$
$$\le C_1 h_k^\kappa\, \Delta\lambda + \eta' h_k^\kappa + C_{2o}(\eta' h_{k-1}^\kappa + \eta h_{k-1}^\kappa) \le [C_1\, \Delta\lambda + \eta' + C_2(\eta+\eta')]\, h_k^\kappa.$$

By (4.5c) one obtains

$$\|u_k^o - u_k(\lambda)\|_U \le \frac{C_1\Delta\lambda + (1+C_2)\eta'}{1-C_2\rho^i}\, h_k^\kappa.$$

Thanks to (4.4b), u_k^o belongs to $\mathcal{U}_k^\lambda(\varepsilon_k)$. Thus, the subsequent iteration by ϕ_k is well-defined and results in $\tilde{u}_k$ with

$$\|\tilde{u}_k - u_k(\lambda)\|_U \le \rho^i \|u_k^o - u_k(\lambda)\|_U \le \frac{C_1\Delta\lambda + (1+C_2)\eta'}{1-C_2\rho^i}\ \rho^i h_k^\kappa = \eta h_k^\kappa$$

(cf. (3.7a,b), (4.5c)) proving (4.6) for k, too. ∎

If the values $\tilde{u}_k$ are used as auxiliary values for the multi-grid iteration (2.13), they have to satisfy (2.12):

Corollary 4.2 The algorithm (4.1) results in $\tilde{u}_k \in \mathcal{U}_k^\lambda(\varepsilon_k/6)$ if in addition to the conditions of Lemma 4.1 the inequality (2.26), $\rho^i \le 1/6$, holds. As in Corollary 2.6 this condition can be weakened to $\eta h_k^\kappa \le (5-\rho)\varepsilon_k / [6(2\rho+1)]$.

Lemma 4.1 shows that it is reasonable to assume the error to behave as $O(h_k^\kappa)$, since the assumption (4.5a) at λ' leads to a similar result at λ. If the nested iteration of § 2 is applied at the first value λ_o, the estimate (4.5a) is established at the starting point. Moving from λ' to λ, we desire $\eta \le \eta'$. The following theorem arises from Lemma 4.1 by setting $\eta = \eta'$.

Theorem 4.3 Suppose

- (3.7a,b) [ϕ_k have contraction number ρ for $u_k \in \mathcal{U}_k^\lambda(\varepsilon_k)$],
- (2.23a-c) [p and h_{k-1} / h_k bounded by C_{2o}, C_{21}; $C_2 = C_{2o}C_{21}^\kappa$],
- (4.3) [discretization error of $\partial u_k / \partial\lambda$]

and choose i and $\Delta\lambda$ ($= \lambda_{\nu+1} - \lambda_\nu$ for all ν) such that

(4.7a) $\rho^i(1+2C_2) < 1$,

(4.7b) $C_1\Delta\lambda h_k^\kappa / [1-\rho^i(1+2C_2)] \le \varepsilon_k$ $(1 \le k \le \ell)$.

Assume that the errors at $\lambda' = \lambda_\nu$ be bounded by

(4.8a) $$\|\tilde{u}_k' - u_k(\lambda')\|_U \le \frac{\rho^i}{1-\rho^i(1+2C_2)}\ C_1\Delta\lambda h_k^\kappa \quad (0 \le k \le \ell).$$

and that the error of $\tilde{u}_o$ at $\lambda = \lambda_{\nu+1}$ is

(4.8b) $$\|\tilde{u}_o - u_o(\lambda)\|_U \le \frac{\rho^i}{1-\rho^i(1+2C_2)}\ C_1\Delta\lambda h_o^\kappa.$$

Then the nested iteration (4.1) is well-defined and produces approximations $\tilde{u}_k$ at $\lambda = \lambda_{\nu+1}$ with the same estimates as in (4.8a):

(4.9) $$\|\tilde{u}_k - u_k(\lambda)\|_U \le \frac{\rho^i}{1-\rho^i(1+2C_2)}\ C_1\Delta\lambda h_k^\kappa .$$

Proof. Set $\eta' = \rho^i C_1\Delta\lambda / [1-\rho^i(1+2C_2)]$. Lemma 4.1 yields (4.6). Since η from (4.5c) coincides with η', (4.9) is proved. ∎

Theorem 4.3 applies to one step from λ_ν to $\lambda_{\nu+1}$. Obviously, we obtain

Corollary 4.4 Suppose (3.7), (2.23), (4.7), and $\lambda_\mu = \lambda_o + \mu\Delta\lambda$ $(0 \le \mu \le N)$. Assume (4.8a) at $\lambda' = \lambda_o$ and (4.8b) for all λ_μ. Then (4.9) is true for all $\lambda = \lambda_\mu$ $(0 \le \mu \le N)$.

For the multi-grid iteration (2.13) the following corollary is important (cf. Corollary 4.2):

Corollary 4.5 The algorithm (4.1) results in $\tilde{u}_k \in \mathcal{U}_k^\lambda(\varepsilon_k/6)$ if in addition to the suppositions of Theorem 4.3 the inequality (2.26), $\rho^i \le 1/6$, holds. If $C_2 \ge 2.5$, (2.26) follows from (4.7a). Otherwise, the condition can be weakened to $6(2\not{p}+1)\ C_1\rho^i\Delta\lambda h_k^\kappa\ /\ [\,(1-\rho^i(1+2C_2))(5-\rho)] \le \varepsilon_k$.

The strategy for the continuation problem is now as follows:

a) Choose η such that $\eta h_k^\kappa \le \varepsilon_k$ $(1 \le k \le \ell)$ and that ηh_k^κ is an acceptable error of $u_k(\lambda)$.

b) Choose i such that, e.g., $\rho^i(1+2C_2) \le 1/2$. Note that C_2 is well-known (cf. Note 2.4).

c) Choose $\Delta\lambda$ sufficiently small so that $2C_1\ \Delta\lambda\ \rho^i \le \eta$. For the estimation of C_1 we refer to § 4.3.1.

It is plain that the claims (i,iii,iv) of § 4.1 are met. Moreover, claim (ii) applies to the nested iteration (4.1). Assuming the standard case $C_2 = 2^2 = 4$ $(\kappa=2)$, we find that i=1 iteration suffices for rates $\rho \le 0.1$, while i=2 iterations are sufficient for $\rho \le 0.3$. However, it must be admitted that the condition $\rho^i(1+2C_2) < 1$ of algorithm (4.1) is more restrictive than the inequality $\rho^i C_2 < 1$ in case of the nested iteration (2.20).

Note that the nested iteration (4.1) can be applied even if the discretization error (4.3) is not $O(\Delta\lambda h_k^\kappa)$ but only $O(h_k^\kappa)$, provided that (4.7b) holds. That means, even if $u_k(\lambda)$ behaves very unsmooth with respect to λ, the new nested iteration is still applicable and similarly effective as the nested iteration (2.20).

The total computational work of the nested iteration (4.1) applied to a fixed interval I is proportional to $i/\Delta\lambda$. To get a fixed accuracy ηh_k^κ, one has to choose $\Delta\lambda$ proportional to ρ^{-i}. Thus, the work is $O(i\rho^i)$. Assume $\rho \le 1/2$. Then $i\rho^i$ decreases for $i \to \infty$ [equivalent to $\Delta\lambda \to \infty$], whence we obtain that one should choose $\Delta\lambda$ as large as possible: Either $\Delta\lambda$ is the a priori given fineness of the λ-partition or $\Delta\lambda$ is the maximal width satisfying (4.7b). We summarize:

Note 4.6 Let $\rho \le 1/2$. Then the computational work (leading to a fixed accuracy) cannot be reduced by decreasing $\Delta\lambda$ and i.

4.3 Additional Comments

4.3.1 Estimation of the Error

As remarked in Note 2.4 and Note 2.6, the constand C_2 is well-known and the contraction number ρ can be observed in course of the computation. To evaluate the right-hand side of the error estimate (4.9), one needs $C_1 \Delta\lambda h_k^\kappa$. Also the constant C_1 can roughly be approximated by the maximum of the values $\|\tilde{u}_k(\lambda) - \tilde{u}_k(\lambda') - \tilde{p}[\tilde{u}_{k-1}(\lambda) - \tilde{u}_{k-1}(\lambda')]\|_U / [h_k^\kappa \Delta\lambda]$ appearing during the computation.

If the approximations $\tilde{u}_k(\lambda)$, $\tilde{u}_k(\lambda')$ etc. satisfy (4.9) and if $\rho^i(3+4C_2) < 1$, then

$$\tilde{C}_1 = \frac{1-\rho^i(1+C_2)}{1-\rho^i(3+4C_2)} \max \{\|\tilde{u}_k(\lambda) - \tilde{u}_k(\lambda') - \tilde{p}[\tilde{u}_{k-1}(\lambda) - \tilde{u}_{k-1}(\lambda')]\|_U/[h_k^\kappa \Delta\lambda]\}$$

is a rigorous upper bound for C_1.

4.3.2 Non-Equidistant λ-Steps

The error estimate of Theorem 4.3 hold for the equidistant choice of $\lambda_\mu = \lambda_o+\mu\Delta\lambda$. However, it might be reasonable to use different λ-steps. Can we still expect the error at $\lambda_{\nu+1}$ to be $\|\tilde{u}_k - u_k(\lambda_{\nu+1})\|_U = O(h_k^\kappa(\lambda_{\nu+1}-\lambda_\nu))$? From Lemma 4.1 it can be seen that this estimate is not true if $\lambda_{\nu+1} - \lambda_\nu \ll \lambda_\nu - \lambda_{\nu-1}$. But if the step length it not arbitrarily decreased, an error $O(h_k^\kappa(\lambda_{\nu+1}-\lambda_\nu))$ can be guaranteed:

Theorem 4.7 Suppose (3.7a,b), (2.22a-c), (4.3), and (4.7a). Choose some $\beta \in (0,1]$ such that

$$\beta > (1+C_2)\rho^i / (1-C_2\rho^i)$$

and define the maximal step length by

$$\Delta\max = \max_{1\le k\le \ell} \varepsilon_k[1-C_2\rho^i - \frac{(1+C_2)\rho^i}{\beta}] / [C_1 h_k^\kappa(1-C_2\rho^i)].$$

The λ-steps have to be chosen so that

$$\beta(\lambda_\nu-\lambda_{\nu-1}) \le \lambda_{\nu+1} - \lambda_\nu \le \Delta\max,$$

i.e., the ratio $(\lambda_{\nu+1}-\lambda_\nu)/(\lambda_\nu-\lambda_{\nu-1})$ is bounded from below by $\beta > 0$. If the errors at the first λ-value λ_o and at level k=0 satisfy

$$\|\tilde{u}_k(\lambda_o)-u_k(\lambda_o)\|_U \le \frac{\Theta}{\beta}(\lambda_1-\lambda_o)h_k^\kappa, \quad \|\tilde{u}_o(\lambda_\mu)-u_o(\lambda_\mu)\|_U \le \Theta(\lambda_\mu-\lambda_{\mu-1})h_o^\kappa \quad (1\le\mu\le\nu)$$

with $\Theta = C_1\rho^i / [1-C_2\rho^i-(1+C_2)\rho^i/\beta]$, then (4.10) holds:

$$(4.10) \qquad \|\tilde{u}_k(\lambda_\nu) - u_k(\lambda_\nu)\|_U \le \Theta(\lambda_\nu-\lambda_{\nu-1})h_k^\kappa.$$

Proof. Assume (4.10) for ν-1 instead of ν. The condition on $\lambda_\nu-\lambda_{\nu-1}$ implies $\|\tilde{u}_k(\lambda_{\nu-1}) - u_k(\lambda_{\nu-1})\|_U \le \frac{\Theta}{\beta}(\lambda_\nu-\lambda_{\nu-1})h_k^\kappa$. For ν=1 this estimate is directly supposed. Thus, inequality (4.5a) of Lemma 4.1 holds for $\lambda' = \lambda_{\nu-1}$ and $\eta' = \Theta(\lambda_\nu-\lambda_{\nu-1}) / \beta$. Also condition (4.4b) is fulfilled since $\lambda_\nu-\lambda_{\nu-1} \le \Delta\max$. By the estimate of $\tilde{u}_o(\lambda_\nu) - u_o(\lambda_\nu)$, the last condition (4.5b) is satisfied, too, and Lemma 4.1 implies $\|\tilde{u}_k(\lambda_\nu) - u_k(\lambda_\nu)\|_U \le \mu h_k^\nu$ with $\eta = [(1+C_2)\eta' + C_1(\lambda_\nu-\lambda_{\nu-1})]\rho^i / [1-C_2\rho^i] = \Theta(\lambda_\nu-\lambda_{\nu-1})$ by definition of Θ. Hence, (4.10) is proved. ∎

4.3.3 Possible Saving of Computational Work

In claim (i) of §4.1 we required that $u_k(\lambda)$ is to be computed for all levels

$k=1,\ldots,\ell$. However, the following situation can arise: We desire approximations $u_k(\lambda)$ $(1\leq k\leq\ell)$ at $\lambda_\nu = \lambda_o + \nu\Delta\lambda'$. But the given step length $\Delta\lambda'$ is too large for satisfying condition (4.7b): $C\,\Delta\lambda' h_k^\kappa \leq \varepsilon_k$ $(1\leq k\leq\ell)$ with $C = C_1 / [1-\rho^i(1+2C_2)]$. Consider the following example. Let

$$\varepsilon_k = \varepsilon h_k,\ \kappa=2,\ h_k = h_{k-1} / 2.$$

Then inequality (4.7b) becomes $\Delta\lambda' \leq \frac{\varepsilon}{C} h_k^{-1}$ $(1\leq k\leq\ell)$. Assume that

$$\Delta\lambda' \leq \frac{\varepsilon}{C} h_2^{-1} \qquad (C = C_1/[1-\rho^i(1+2C_2)])$$

but $\Delta\lambda' > \frac{\varepsilon}{C} h_1^{-1}$, i.e., (4.7b) is fulfilled for $k\geq 2$ but violated for $k=1$. By $h_o = h_1/2$, the step width $\Delta\lambda := \Delta\lambda'/2$ satisfies (4.7b) for $k=1$, too. In this case one can proceed as follows (cf. Fig. 7) :

- for all $\lambda_\nu=\nu\Delta\lambda$ apply the nested iteration (4.1) with 1 instead of ℓ. As usual the values $\tilde{u}_k'$ are those at $\lambda_{\nu-1}$.

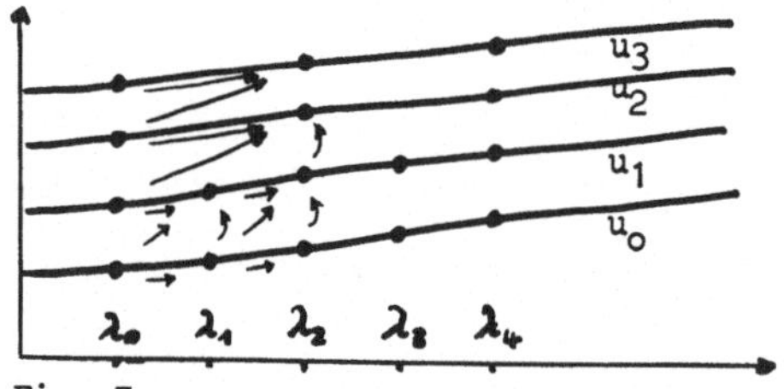

Fig. 7

- for all $\lambda_\nu=\lambda_o+\nu\Delta\lambda=\lambda_o+\frac{\nu}{2}\Delta\lambda'$ with ν even, apply the nested iteration (4.1) starting at k=1 (instead of k=0) with step width $\Delta\lambda'$ and $\tilde{u}_m'=\tilde{u}_m(\lambda_{\nu-2})$.

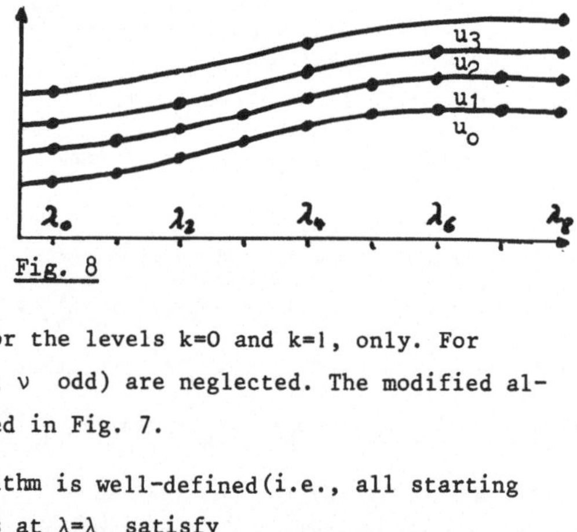

Fig. 8

Thus, the smaller step width $\Delta\lambda$ is used for the levels k=0 and k=1, only. For $k\geq 2$ the intermediate values $\tilde{u}_k(\lambda_\nu)$ $(k=0,1;\ \nu$ odd) are neglected. The modified algorithm results in approximations indicated in Fig. 7.

It is easy to verify that the algorithm is well-defined (i.e., all starting values are in $\mathcal{U}_k^\lambda(\varepsilon_k)$) and that the results at $\lambda=\lambda_\nu$ satisfy

$$(4.11)\qquad \|\tilde{u}_k(\lambda_\nu)-u_k(\lambda_\nu)\|_U \leq \frac{\rho^i}{1-\rho^i(1+2C_2)}\, C_1 h_k^\kappa\, \Delta\lambda_k \text{ with } \Delta\lambda_k = \begin{cases}\Delta\lambda' & (k>1)\\ \Delta\lambda=\Delta\lambda'/2 & (k\leq 1)\end{cases}$$

for $0\leq k\leq 1$ if ν odd and for $0\leq k\leq\ell$ if ν even.

It is obvious how to extend the modification to step widths $\Delta\lambda_k = \Delta\lambda\cdot 2^k$ different for all k. This approach leads to the results depicted in Fig. 8. Then condition (4.7b) becomes

$$C_1\ \Delta\lambda_k\ h_k^{\kappa}\ /\ [1-\rho^{i}(1+2C_2)] \le \varepsilon_k,$$

whereas the approximations satisfy the error estimate

$$\|\tilde{u}_k(\lambda_\nu) - u_k(\lambda_\nu)\|_U \le \frac{\rho^{i}}{1-\rho^{i}(1+2C_2)}\ C_1\ \Delta\lambda_k\ h_k^{\kappa}$$

instead of (4.8b).

The computational work of the described algorithm is proportional to $1/\Delta\lambda_\ell$. Let W_k be the work of one iteration ϕ_k. The work of the continuation from $\lambda_o + \mu\Delta\lambda_\ell$ to $\lambda_o + (\mu+1)\Delta\lambda_\ell$ requires $i[W_\ell + 2W_{\ell-1} + \ldots + 2^{\ell}W_o]$. In the two-dimensional case, the relation $W_k \approx 4W_{k-1}$ shows that the total work is bounded by $2iW_\ell$. Thus, the computational work for this iteration is not much larger than the nested iteration (4.1) with $\Delta\lambda = \Delta\lambda_\ell$ taking the work $\le \frac{4}{3}\, iW_\ell$.

4.3.4 Extrapolation

The excellently small errors of $\tilde{u}_k(\lambda)$ enable us to apply the Richardson extrapolation of $u_k(\lambda) = u_{h_k}(\lambda)$ to the limit $h_k \to 0$. The extrapolated value

$$\tilde{u}_{ex}(\lambda) := [\tilde{u}_k(\lambda) - (h_k/h_{k-1})^{\kappa}\ \tilde{u}_{k-1}(\lambda)]\ /[1-(h_k/h_{k-1})^{\kappa}]$$

approximates the continuous solution $u(\lambda)$ up to a term $O(h_k^{\kappa}(h_k^{\tau}+\Delta\lambda))$ $(\tau > 0)$, provided that $u_k(\lambda)$ admits the asymptotic expansion $u_k(\lambda) = u(\lambda) + h_k^{\kappa}\ w(\lambda) + O(h_k^{\kappa+\tau})$ with $w(\lambda)$ independent of h_k.

4.3.5 Frozen-τ Technique

For continuation methods Brandt [5] recommends the 'frozen-τ technique'. This approach is based on the fact that $\hat{u}_k(\lambda) := \hat{r}_{k\ell}u_\ell(\lambda)$ ($\hat{r}_{k\ell}$: restriction from level ℓ to k; e.g. $\hat{r}_{k\ell}$ = trivial injection onto the coarse grid) is the solution of

$$L_k(\hat{u}_k(\lambda),\lambda) = \tau_{k\ell}(\lambda) \qquad \text{with } \tau_{k\ell}(\lambda) := L_k(\hat{r}_{k\ell}u_\ell(\lambda),\lambda).$$

By definition, the discretization error of $\hat{u}_k$ is h_ℓ^{κ} (not h_k^{κ}). Assume that u_ℓ is known at λ_ν; hence, $\tau_{k\ell}(\lambda_\nu)$ can be computed. Replacing $\tau_{k\ell}(\lambda)$ by the 'frozen' value $\tau_{k\ell}(\lambda_\nu)$ one tries to approximate $\hat{u}_k(\lambda)$ by the solution $\hat{u}_k^{fr}(\lambda)$ of $L_k(\hat{u}_k^{fr}(\lambda),\lambda) = \tau_{k\ell}(\lambda_\nu)$.

Under suitable smoothness conditions it can be shown that $\partial\tau_{k\ell}(\lambda)\ /\ \partial\lambda$ is of order $O(h_k^{\kappa})$ and thereby

$$\|\hat{u}_k^{fr}(\lambda) - \hat{u}_k(\lambda)\|_U \le \hat{C}_1 |\lambda-\lambda_\nu| h_k^{\kappa}\ .$$

Since we have only an approximation $\tilde{u}_\ell(\lambda_\nu)$ instead of $u_\ell(\lambda_\nu)$, this approach yields $\tilde{\hat{u}}_k^{fr}(\lambda)$ with

$$(4.12)\qquad \|\tilde{\hat{u}}_k^{fr}(\lambda) - \hat{u}_k(\lambda)\|_U \le \hat{C}_1 |\lambda-\lambda_\nu| h_k^{\kappa} + C_o\|\tilde{u}_\ell(\lambda_\nu) - u_\ell(\lambda_\nu)\|_U.$$

Notice that $\hat{u}_k(\lambda)$ is a replacement for $u_\ell(\lambda)$. According to claim (iii), namely that the errors of $\tilde{\hat{u}}_k$ should be of the same order as the discretization error $O(h_\ell^{\kappa})$ (not $O(h_k^{\kappa})$!), we have either

- to reiterate at level ℓ with starting value $u_\ell^o(\lambda_{\nu+1}) = p\tilde{\hat{u}}_k^{fr}(\lambda_{\nu+1})$ in order to reduce the error of $\tilde{u}_\ell(\lambda_{\nu+1})$ to $O(h_\ell^\kappa)$

  or

- to restrict the frozen-τ technique to the interval $[\lambda_\nu,\lambda_{\nu+1}]$ with length $\lambda_{\nu+1} - \lambda_\nu = O(h_\ell^\kappa/h_k^\kappa)$ and then to restart as in the first case.

<u>First strategy</u>. Choosing the step width $\Delta\lambda$ (cf. Note 4.6) and assuming that the starting iterate u_ℓ^o belongs to $U_\ell^\lambda(\varepsilon_\ell)$, we obtain error of $\tilde{u}_\ell$ bounded by

$$\hat{C}_1(h_k/h_\ell)^\kappa(\rho^i\Delta\lambda h_\ell^\kappa) \;/\; [1-\rho^i(C_2+C_o)].$$

By a comparable amount of work the nested iteration (4.1) yields results with error

$$C_1(\rho^i\Delta\lambda h_\ell^\kappa) \;/\; [1-\rho^i(1+2C_2)].$$

Thus, we have to compare $\hat{C}_1(h_k/h_\ell)^\kappa$ and C_1. Obviously, $(h_k/h_\ell)^\kappa$ [usually equal to $2^{\kappa(\ell-k)}$] is >> 1. For the comparison of $\hat{C}_1$ and C_1 we refer to

<u>Note 4.8</u> Let 2m be the order of the differential operator $L(.,\lambda)$. Then $\hat{C}_1$ depends on the $(2m+\kappa)^{th}$ (spatial) differences of $\partial u_\ell(\lambda)/\partial\lambda$, whereas C_1 depends on the κth (spatial) differences. Therefore, $\hat{C}_1$ can be much larger than C_1.

<u>Second strategy</u>. Let $\Delta\lambda' = N\cdot\Delta\lambda$, $\lambda_\nu = \lambda_o + \nu\Delta\lambda$, and restart for $\lambda = \lambda_{\mu N}$ ($\mu \in \mathbb{N}$). The errors are of quite another kind than in the previous cases. At $\lambda = \lambda_{\mu N}$ the error can be controlled by means of i(= number of ϕ_ℓ-iterations). However, for all other λ_ν the error exceeds

$$\hat{C}_1|\lambda_\nu - \lambda_{\mu N}|\,h_k^\kappa$$

independently of i. The maximum of these terms is about $\hat{C}_1\Delta\lambda' h_k^\kappa$. E.g., suppose $k = \ell-1$. We have to solve one equation $L_{\ell-1}(u_{\ell-1},\lambda_\nu) = \tau_{\ell-1,\ell}(\lambda_{\mu N})$ for each λ_ν. In the two-dimensional case this work corresponds to $\frac{1}{4} W_\ell(\Delta\lambda)$, where $W_\ell(\Delta\lambda)$ = work for solving N problems at level ℓ at each λ_ν, $\mu N < \nu \leq (\mu+1)N$. Note that the total work of the second strategy is of course larger than $\frac{1}{4} W_\ell(\Delta\lambda)$ because of the restarts.

To compare this algorithm with the nested iteration (4.1) consider the following modification: Apply the nested iteration (4.1) with $i \geq 1$ for $\nu \equiv 0 \pmod 4$ and $i = 0$, otherwise. Then the total work is $\frac{1}{4} W_\ell(\Delta\lambda)$, too. But even at the intermediate values λ_ν the error improves with ρ^i, whereas the error (4.12) of the second frozen-τ version does not. Therefore, we conclude that the nested iteration (4.1) is superior to the frozen-τ technique.

5. Treatment of Turning Points

5.1 Rescription of the Problem

Let us consider the situation of Fig. 2. A possible picture of the discrete solutions is sketched in Fig. 9. There are two

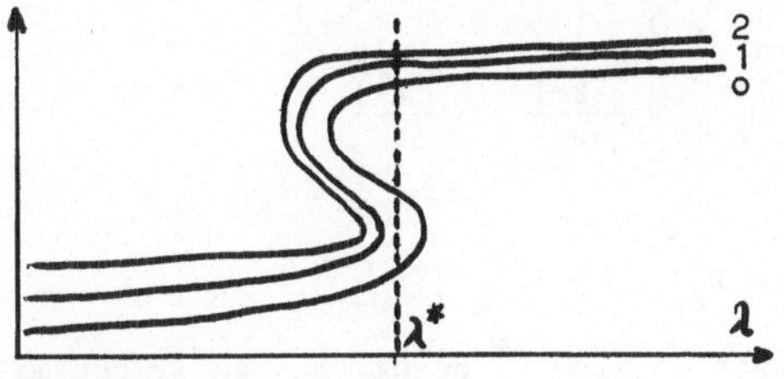

Fig. 9

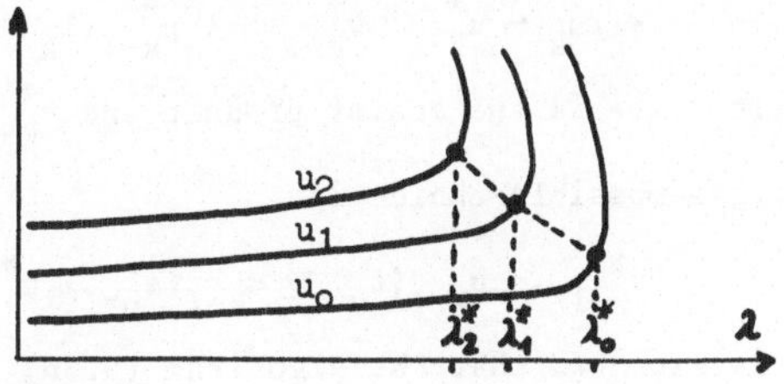

Fig. 1o

closely related problems arising from the existence of a turning point. First, the continuous problems (as well as the discrete ours) become increasingly ill-posed when λ approaches the turning point (cf. Chan and Keller [7]). Second, solutions of a level k might have no counterpart at another level. E.g., in Fig. 9 there are two solutions $u_o(\lambda^*)$ in the lower part of the branch related to no solution $u_1(\lambda^*)$ at level 1.

5.2 Change of Parameter

A very simple remedy is the introduction of a new parameter t such that turning points vanish in the t-diagram. E.g., in Fig. 9 the ordinate can serve as t.

If again the reformulated problem can be written as $\hat{L}_k(u_k(t),t) = 0$, we regain the situation of §§ 2 - 4. But often it may happen that one obtains a new problem for $u_k(t)$ *and* $\lambda_k(t)$ (now depending on k!) described by

$$(5.1) \qquad L_k(u_k(t),\lambda_k(t)) = 0, \qquad \psi_k(u_k(t),t) = 0.$$

If, e.g., $t = \Psi_k(u_k) \in \mathbb{R}$ ist the ordinate in Fig. 9, the (scalar) function ψ, is defined by $\psi_k(u_k,t) = \Psi_k(u_k) - t$. An example for a modification of (5.1) is given in the following subsection.

5.3 Arc-Length Continuation

The arc-length s is defined by

$$(5.2) \qquad \|\frac{d}{ds}\, u(s)\|_u^2 + |\frac{d}{ds}\, \lambda(s)|^2 = 1$$

(cf. Keller [15]). A first approach would be as follows: Start at some λ_o with $u_k(\lambda_o)$ assumed to be well-defined. Set $u_{k|\,s=o} = u_k(\lambda_o)$, $\lambda(o) = \lambda_o$ and try to solve

$$L_k(u_k(s_\nu), \lambda_k(s_\nu) = 0,$$
$$\|u_k(s_\nu) - u_k(s_{\nu-1})\|_U^2 + |\lambda_k(s_\nu) - \lambda_k(s_{\nu-1})|^2 = \Delta s^2.$$

A possibly better approach is to use arc-length continuation only at the lowest level 0, combined with the condition that $ru_k - u_{k-1}$ is perpendicular to the branch $u_{k-1}(s)$. More precisely, for given $u_{k-1}^* = u_{k-1}(s_\nu)$ and $\lambda_{k-1}^* = \lambda_{k-1}(s_\nu)$ the values u_k^* and λ_k^* are defined as solutions of

$$(5.3a) \qquad L_k(u_k,\lambda_k) = 0;$$

(5.3b) $\quad < ru_k - u^*_{k-1},\ v^*_{k-1} > + \mu^*_{k-1}(\lambda_k - \lambda^*_{k-1}) = 0,$

where <.,.> is the scalar product and $v^*_{k-1} : \mu^*_{k-1} \approx \frac{\partial u_{k-1}}{\partial s} : \frac{\partial \lambda_{k-1}}{\partial s}$.

E.g., a possible choice is

$$v^*_{k-1} := u_{k-1}(s_\nu) - u_{k-1}(s_{\nu-1}), \mu^*_{k-1} := \lambda_{k-1}(s_\nu) - \lambda_{k-1}(s_{\nu-1}).$$

It is expected that the algorithm (5.3b) selects a value u^*_k being the nearly optimal approximation of u^*_{k-1} at level k.

5.4 Discrete Elliptic Scheme Combined with an Additional Scalar Equation

In contrast to the situation in §§ 2 - 4 we have to solve a system of n_k equations

(5.4a) $\quad L_k(u_k,\lambda_k) = 0$

and a further *scalar* equation of the type

(5.4b) $\quad \Lambda_k(u_k,\lambda_k) = 0.$

In case of Eq. (5.1), Λ_k is defined by $\Lambda_k(u_k,\lambda_k) = \psi_k(u_k,t)$ for fixed t, while in case of Eq. (5.3b) we have $\Lambda_k(u_k,\lambda_k) = < ru_k - u^*_{k-1},\ v^*_{k-1} > + \mu^*_{k-1}(\lambda_k-\lambda^*_{k-1})$ for fixed $u^*_{k-1}, v^*_{k-1}, \mu^*_{k-1}, \lambda^*_{k-1}$. Note that these 'fixed' values are computed by the nested iteration (4.1) at level k-1 before the computation starts at level k.

In § 2 we described the nonlinear multi-grid iteration of the system $L_k(u_k,\lambda_k) = 0$ with fixed λ_k. Now we have to explain the multi-grid iteration for the extended system (5.4a,b) and its inhomogeneous version

(5.5) $\quad L_k(u_k,\lambda_k) = f_k, \quad \Lambda_k(u_k,\lambda_k) = \alpha_k.$

It is formally the same algorithm (2.13) with the following replacements:

- smoothing in step (2.13ba,bf): Apply the original smoothing for $L_k(u_k,\lambda_k) = f_k$ with *fixed* λ_k. Thus, there is no smoothing with respect to the component λ_k. λ_k remains unchanged: $\lambda_k' = \lambda_k$.
- the defect d_k in step (2.13bb) consists now of n_k components $L_k(u_k',\lambda_k') - f_k$ and a $(n_k+1)^{st}$ component $\Lambda_k(u_k',\lambda_k') - \alpha_k$.
- restriction: in step (2.13bd) the restriction rd_k has to be defined. The first n_{k-1} components of rd_k form the usual restriction of the first n_k components of d_k. The last component of rd_k coincides with the last component of d_k.
- prolongation in step (2.13be): the first n_k components of pw_{k-1} form the usual prolongation of the first n_{k-1} components of w_{k-1}. The last component of pw_{k-1} equals the last component of w_{k-1}.

Obviously, the performance of the modified algorithm is as easy as the original one, and it requires the same subroutines for smoothing, restriction, and prolongation.

Note that the smoothing procedure for the extended system does not converge to the solution of (5.5) since λ_k is unchanged. However, the solution of (5.5) is a fixed point of the extended smoothing iteration. Errors arising from wrong λ_k values can be regarded as smooth components that are to be eliminated by the coarse-grid correction, not by the smoothing step. Nonetheless, to correct λ_k, one may add the following correction after the smoothing step:

$$u_k \to u_k + \omega v_k^*, \quad \lambda_k \to \lambda_k + \omega\mu_k^*$$

with ω from $\Lambda_k(u_k + \omega v_k, \lambda_k + \omega\mu_k^*) \approx \alpha_k$ and $v_k^* : \mu_k^* \approx \frac{d}{ds} u_k(s) : \frac{d}{ds} \lambda_k(s)$.

A comment on the convergence of the extended multi-grid iteration is given in

Note 5.1 Let (u_k^*,λ_k^*) be the solution of (5.4a,b). We express the relative consistency of Λ_k and Λ_{k-1} by

$$(5.6)\quad \frac{\partial}{\partial u_k}\Lambda_k(u_k^*,\lambda_k^*)\cdot p - \frac{\partial}{\partial u_{k-1}}\Lambda_{k-1}(u_{k-1}^*,\lambda_{k-1}^*) = O(h_k^\kappa),\quad \frac{\partial}{\partial\lambda_k}\Lambda_k(\ldots) - \frac{\partial}{\partial\lambda_{k-1}}\Lambda_{k-1}(\ldots) = O(h_k^\kappa)$$

with $\kappa > 0$. Then it is possible to show that the modified multi-grid iteration converges under usual conditions, provided that h_o is sufficiently small.
The last condition is awkward. But often, the difficulties can be removed by the following construction.

Assume that the approximations $(u_o^*,\lambda_o^*),\ldots,(u_{k-1}^*,\lambda_{k-1}^*)$ of Eq. (5.4a,b) are already computed. In order to solve Eq. (5.4a,b) at level k, $L_k(u_k^*,\lambda_k^*) = 0$, $\Lambda_k(u_k^*,\lambda_k^*) = 0$, we modify the Λ_k-equations at the lower levels:

$$(5.7a)\quad L_m(u_m,\lambda_m) = f_m \qquad (0 \le m \le k-1),$$

$$(5.7b)\quad \Lambda_m^k(u_m,\lambda_m) := \Lambda_m(u_m^*,\lambda_m^*) + \langle w_m^k, u_m - u_m^*\rangle + \omega_m^k(\lambda_m - \lambda_m^*) = \alpha_m \quad (0 \le m \le k-1),$$

where $\omega_m^k \approx \partial\Lambda_k(u_k^*,\lambda_k^*)/\partial\lambda_k$, $w_m^k \approx (\partial\Lambda_k(u_k^*,\lambda_k^*)/\partial u_k)p_{km}$, $p_{km} = p_{k,k-1}p_{k-1,k-2}\cdots p_{m+1,m}$ and $p_{q,q-1} = p$ ist the usual prolongation from level q-1 to q.
Again, (u_m^*,λ_m^*) is an approximation of the new problem (5.7a,b). Note that by our construction the terms of (5.6) vanish.

We give an example for the affine function

$$\Lambda_k(u_k,\lambda_k) = \langle r'u_k - u_{k-1}^*,\ v_{k-1}^*\rangle + \mu_{k-1}^*(\lambda_k - \lambda_{k-1}^*)$$

from (5.3b). The new Λ-functions are

$$\Lambda_m^k(u_m,\lambda_m) = c_m + \langle r'p_{km}u_m, v_{k-1}^*\rangle + \mu_{k-1}^*\lambda_m \qquad (0 \le m \le k-1)$$

with constants c_m such that $\Lambda_m^k(u_m^*,\lambda_m^*) = 0$. A possible choice of r' is r'=trivial injection onto the coarser grid. Then for any interpolation $p_{k,k-1} = p$ the relation $r'p = I$ implies $\langle r'p_{km}u_m, v_{k-1}^*\rangle = \langle p_{k-1,m}u_m, v_{k-1}^*\rangle$. Defining the adjoint mapping of $p_{k-1,m}$ by $r_{m,k-1} = p_{k-1,m}^*$, we obtain

$$\Lambda_m^k(u_m,\lambda_m) = c_m + \langle u_m, v_m^{**}\rangle + \mu_{k-1}^*\lambda_m$$

with $v_m^{**} = r_{m,k-1} v_{k-1}^{*}$.

By this construction we overcome the problem remarked above:

Note 5.2 If Λ_k is an affine mapping and if the equations at the lower levels are defined by (5.7a,b), then the multi-grid iteration extended as described above (but without smoothing step (2.13bf) after the coarse-grid correction) yields iterates (u_k^j, λ_k^j) satisfying $\Lambda_k(u_k^j, \lambda_k^j) = \alpha_k$ exactly. Here it is assumed that the results at level m=0 fulfil the respective Λ_o^k-equation.

Proof. Formally, we set $\Lambda_k^k = \Lambda_k$ and prove Note 5.2 induction over $m = 0,\dots,k$. By assumption the assertion holds for m=0. Let (u_m', λ_m') be the result after the smoothing step (2.13ba) having the defect $\delta_m = \Lambda_m^k(u_m, \lambda_m) - \alpha_m$. The coarse-grid system contains the equation $\Lambda_{m-1}^k(u_{m-1}, \lambda_{m-1}) = \alpha_{m-1}$ with $\alpha_{m-1} = \sigma\delta_m$ (σ from (2.13bb)). By definition of Λ_{m-1}^k and by the inductive assumption on the solution (u_{m-1}, λ_{m-1}) the differences $\Delta u_{m-1} = (u_{m-1} - u_{m-1}^{*}) / \sigma$ and $\Delta\lambda_{m-1} = (\lambda_{m-1} - \lambda_{m-1}^{*}) / \sigma$ satisfy

$$\frac{\partial}{\partial u_{m-1}} \Lambda_{m-1}^k \Delta u_{m-1} + \frac{\partial}{\partial \lambda_{m-1}} \Lambda_{m-1}^k \Delta\lambda_{m-1} = \alpha_{m-1} / \sigma = \delta_m .$$

Consequently, the coarse-grid correction $(u_m', \lambda_m') \to (u_m' - p\Delta u_m,\ \lambda_m' - \Delta\lambda_m)$ yields a new iterate satisfying

$$\Lambda_m^k(u_m^j, \lambda_m^j) - \alpha_m = \Lambda_m^k(u_m' - \Delta u_m, \lambda_m' - \Delta\lambda_m) - \alpha_m = \delta_m - \frac{\partial \Lambda_m^k}{\partial u_m} p\Delta u_{m-1} -$$
$$- \frac{\partial \Lambda_m^k}{\partial \lambda_m} \Delta\lambda_{m-1} = \delta_m - \frac{\delta \Lambda_{m-1}^k}{\partial u_{m-1}} \Delta u_{m-1} - \frac{\partial \Lambda_{m-1}^k}{\partial \lambda_{m-1}} = \delta_m - \delta_m = 0. \quad \blacksquare$$

In the nonlinear case the Λ-equation is not completely satisfied, but the error vanishes for $u_m^j \to u_m^{*}$.

We conclude that especially for the choice (5.7a,b) the extended equation (5.4a,b) is solved by the modified multi-grid iteration as easily as the original problem (3.3) in the regular case of §§ 3 - 4. Again, the nested iteration (4.1) can be applied the continuation problems with turning points.

References

1. BANK, R.: A Multi-Level Iterativ Method for Nonlinear Elliptic Equations. In: [17], 1981, pp. 1-16

2. BANK, R.E. and ROSE, D.J.: Analysis of a multilevel iterative method for nonlinear finite element equations. Techn.Rep. 202, 1981

3. BIGGE, J. and BOHL, E.: On the steady states of finitely many chemical cells. To appear

4. BRANDT, A.: Multi-level adaptive solutions to boundary value problems. Math. Comp. 31 (1977), 333-390

5. BRANDT, A.: Multigrid solvers on parallel computers. In: [17], 1981, pp. 39 - 63

6. BREZZI, F., RAPPAZ, J. and RAVIART, P.A.: Finite dimensional approximation of nonlinear problems. I: Numer.Math. 36 (1981), 1 - 25; II: Numer.Math. 37(1981), 1 - 28; III: Numer.Math. 38 (1981), 1 - 30

7. CHAN, T.F.C. and KELLER, H.B.: Arc-length continuation and multi-grid techniques for nonlinear elliptic eigenvalue problems. Report, Yale University, 1981

8. DEUFLHARD, P.: A stepsize control for continuation methods and its special application to multiple shooting techniques. Numer.Math. 33 (1979), 115-146

9. HACKBUSCH, W.: On the fast solution of nonlinear elliptic equations. Numer.Math. 32(1979), 83-95

10. HACKBUSCH, W.: On the convergence of multi-grid iterations. Beiträge zur Numer.Math. 9(1981), 213-239

11. HACKBUSCH, W.: Error analysis of the nonlinear multi-grid method of the second kind. Aplikace Matematiky 26(1981), 18-29

12. HACKBUSCH, W.: Regularity of difference schems-Part II: Regularity estimates for linear and nonlinear problems. To appear in Ark.Mat.

13. HACKBUSCH, W.: Multi-grid convergence theory. In: [14]

14. HACKBUSCH, W. and TROTTENBERG, U. (eds.): Multi-Grid Methods-Conference in Cologne-Porz 1981. Lecture Notes in Mathematics, Springer-Verlag, Heidelberg, to appear in 1982

15. KELLER, H.B.: Numerical solution of bifurcation and nonlinear eigenvalue problems. In: Applications of Bifurcation Theory (P.H. Rabinowitz, ed.). Academic Press, New York, 1981, pp. 359-383

16. KRONSJÖ, L.: A note on the "nested iteration" method. BIT 15(1975), 107-110

17. SCHULTZ, M.H. (ed.): Elliptic Problem Solvers. Academic Press, New York, 1981

18. STÜBEN, K. TROTTENBERG, U.: Multi-grid method: Fundamental algorithm, model problem analysis, and applications. In: [14]

A FAST SOLVER FOR NONLINEAR EIGENVALUE PROBLEMS

H. D. Mittelmann
Abteilung Mathematik
Universität Dortmund
Postfach 5o o5 oo
D-46oo Dortmund 5o/FRG

Abstract A numerical method recently proposed by the author is shown to be a very efficient and robust method for the solution of a class of discrete nonlinear eigenvalue problems. In particular it is applied to follow the relevant and the spurious solution curves. Numerical results show that also in the neighbourhood of turning or bifurcation points the work required is considerably less than for usual continuation procedures and that a larger steplength may be chosen. A corresponding multi-grid method is used for following spurious solution branches.

O. Introduction

We consider the class of nonlinear eigenvalue problems which may be interpreted as the necessary conditions for their solutions to be critical points of a functional with respect to the level surfaces of another functional. This class includes many interesting applications and may easily be generalized to cover nonlinear variational inequalities respectively free boundary problems ([4]).

A numerical method originally developed for these latter problems is applied here exclusively to simple nonlinear eigenvalue problems. Typical features of these problems are the occurrence of one or more turning points of the solution curves as well as bifurcation points. The discretizations may exhibit spurious solutions (see, for example, [1] and the literature cited there) and the question arises if these can be computed by algorithms used for the determination of the relevant solution. It may be even more important if algorithms do not lead to these solutions by mistake or how they may be avoided.

The generalized inverse iteration of [4] is shown in the following to have attractive properties with respect to these questions and to be a very efficient algorithm. It is also robust in the sense that when it is used in the fashion of a continuation procedure the steplength may be chosen rather large.

Spurious solutions often do not cause much trouble when computing the relevant solution since they are present only for relatively large parameter values. Multi-grid algorithms, however, simultaneously use several discretizations and it is of interest how they perform in the presence of spurious solutions.

The contents of the next sections are

1. The generalized inverse iteration
2. Following the relevant solution curves
3. Following the spurious solution curves
4. Path following near turning points
5. Path following near bifurcation points
6. How to avoid spurious solutions
7. A multi-grid method for nonlinear eigenvalue problems.

1. The Generalized Inverse Iteration

In the following we introduce the basic algorithm which will be used for the numerical solution of nonlinear eigenvalue problems. This method may be applied directly to the continuous problem (cf. [4]). We assume, however, that this is discretized yielding a finite-dimensional problem of the form

$$(1.1) \qquad \nabla f(x) = \lambda Bx, \lambda \in \mathbb{R},$$

where $f : \mathbb{R}^n \to \mathbb{R}$ is a sufficiently smooth functional and B a symmetric and positive-definite nxn matrix. We thus look for a critical point of f with respect to level surfaces of the functional $g : \mathbb{R}^n \to \mathbb{R}$ given by

$$(1.2) \qquad g(x) = 1/2\ x^T Bx$$

and λ may be interpreted as the corresponding Lagrange multiplier.

The following algorithm was originally developed in [4] for the solution of bifurcation problems for variational inequalities for which it represents one of the few possible approaches and as numerical experience showed a very efficient one. First computational results in [4] indicated, however, that it is also a competitive method for the solution of problems of the type (1.1), for which a series of algorithms has been proposed and applied.

We need some additional notation. For $x = x_k$, the k-th iterate of the algorithm, we denote by $F_k = F(x_k)$ the Hessian matrix of f and by $\lambda_k = \nabla f(x_k)^T x_k / x_k^T B x_k$ the corresponding Lagrange multiplier. Let further

$$(1.3) \qquad \partial S_\rho = \{x \in \mathbb{R}^n \, , \; g(x) = \rho^2/2\}$$

be the level surface of g on which the critical points are to be determined. Finally we define

$$(1.4) \qquad H_k = \begin{bmatrix} F_k - \lambda_k B & -Bx_k \\ -x_k^T B & 0 \end{bmatrix}^{-1}_{nxn}$$

i. e. H_k is the nxn principal submatrix of the inverse of this augmented matrix, provided it is regular.

<u>The algorithm</u>

Let $x_1 \in \partial S_\rho$ be given. For $k = 1,2,\ldots$ set

$$(1.5a) \qquad x_{k+1} = \rho \tilde{x}_{k+1} / \|\tilde{x}_{k+1}\|_B \, , \quad \text{where}$$

$$(1.5b) \qquad \tilde{x}_{k+1} = x_k - H_k \nabla f(x_k)$$

and $\|\cdot\|_B$ is the norm introduced by B.

The following theorem may be proved for algorithm (1.5).

<u>Theorem 1.6</u> Let f in (1.1) be twice Fréchet differentiable in the neighbourhood of a solution (x_o, λ_o) and assume that

$$(1.7) \qquad N(F_o - \lambda_o B) \subset \operatorname{span} \{Bx_o\} \, .$$

Here N(L) denotes the nullspace of the linear operator L.

Then for $x_1 \in \partial S_{\rho_o}$, $\rho_o = \| x_o \|_B$, sufficiently close to x_o the sequence $\{x_k\}$ generated by (1.5) converges quadratically against x_o.

Under the assumption (1.7) the matrix in (1.4) is regular in a neighbourhood of (x_o, λ_o). Hence algorithm (1.5) which may be rewritten as

$$(1.8) \qquad x_{k+1} = \phi(x_k)$$

is well defined and it was shown in [4] that

$$\phi'(x_o) P_{x_o} = o$$

where P_z denotes the orthogonal projector on $\{z\}^{\perp} = \{x \in \mathbb{R}^n,\ x^T B z = o\}$. This proves the theorem.

We note that the assumption (1.7) allows the computation of solutions in turning points but not in bifurcation points.

2. Following the Relevant Solution Curves

For the following of the solution paths for problems of the type (1.1) algorithm (1.5) may obviously be utilized taking ρ as a parameter if only the steps in ρ are taken sufficiently small. It is, however, not obvious which advantages should come from augmenting the matrix $F_k - \lambda_k B$ of Newton's method for (1.1) and thus increasing the dimension of the linear system by one, except, of course, in a turning point x_o where $F_o - \lambda_o B$ becomes singular. We shall not comment on this here but report now the results of some simple numerical experiments.

We want to demonstrate that algorithm (1.5) is indeed a very efficient but also robust method for path following. In order to have any reference method we include the results for Newton's method taking λ or one of the components of x as parameter (see, for example, [2]). For this method we take the latest computed solution as starting value for the next parameter value. Other numerical methods including λ-continuation with Euler predictor step have been compared with algorithm (1.5) in [4].

Solving a problem of the form (1.1) which is a discretization with discretization parameter $h>o$ of a nonlinear eigenvalue problem we have to bear in mind that (1.1) may have spurious or irrelevant solutions which for $h \to o$ do not converge to a solution of the continuous problem but simply disappear. If the nonlinear system

(2.1) $$\nabla f(x) = o$$

has solutions then the spurious solutions occur for λ sufficiently small, while a sufficiently strong growth of $\nabla f(x)$ for $||x|| \to \infty$ in the case $(\nabla f)_i(x) = (\nabla f)_i(x_i)$, $i = 1,\ldots,n$, may yield spurious solutions for all sufficiently large $||x||$. If the left side of (1.1) is any continuous diagonal field then it is the gradient of a suitable functional. Spurious solutions of both type have already been investigated (see, for example,[1] and the literature cited there). We restrict ourselves to the second type.

The example we shall consider in all sections except 5. is

(2.2) $$\mu \nabla f(x) = Bx, \quad \mu \hat{=} \lambda^{-1}, \quad x \in \mathbb{R}^n$$

where $(\nabla f)_i(x) = \exp(x_i/(1+\varepsilon x_i))$, $\varepsilon \geq o$, and

(2.3a) $$B = B_n^{(1)} = (n+1)^2 \begin{bmatrix} 2 & -1 & & & \\ -1 & 2 & -1 & & \\ & -1 & 2 & -1 & \\ & & \ddots & \ddots & -1 \\ & & & -1 & 2 \end{bmatrix}$$

in one space dimension or

(2.3b) $$B = B_n^{(2)} = (B_n^{(1)} \otimes I_n + I_n \otimes B_n^{(1)})$$

in two space dimensions. Here I_n denotes the nxn identity matrix and for simplicity we assume n to be odd.

For varying ε the relevant solution curve exhibits none, one or two turning points, corresponding to a unique respectively two or three different solutions for a given μ. From the results in [1] it is clear that for $\varepsilon = o$ the relevant solution is asymptotically proportional to the $(n+1)/2$-th column of B^{-1} for (2.3a) while there are spurious solutions corresponding to the other n-1 columns of B^{-1} respectively to a linear combination of certain pairs of the columns of this matrix.

Since we give only the number of steps for the algorithms and not, for example, the computer time, we should comment on the solution of the linear systems in each step. For the λ-continuation a system with matrix $F_k - \lambda_k B$ has to be solved which is symmetric and (block-)tridiagonal but in general indefinite; for ρ-continuation with algorithm (1.5) this matrix is augmented according to (1.4) preserving the symmetry, while for continuation along a component of x the dimension is reduced by one but the structure and the symmetry are destroyed. Hence there is no reason why the solution of (1.4) by direct or iterative methods should be considerably more costly than Newton's method.

For the solution of the linear system in each step of algorithm (1.5) we suggest suitable conjugate-gradient methods as, for example, SYMMLQ ([6]).

We give now the results obtained for (2.2), (2.3a) with $\varepsilon = o$, $n = 15$. At first we follow the relevant branch for a given ρ-sequence with algorithm (1.5) and then for the corresponding parameter values with μ-continuation and continuation along x_8. The results in Table 2.1 show that for a relatively small step size the μ-continuation needs about twice as much iterations as ρ-continuation. Continuation along x_8 in the region up to and beyond the turning point is possible but expensive. The numbers in parantheses indicate that an intermediate step had to be used otherwise the number of iterations was even higher. Our experience with all examples reported here and in the sequel was that the number of iterations was independent of n for ρ-continuation with algorithm (1.5), while this was in general not the case for the other continuation procedures.

A continuation method may be called robust if the step size may be chosen relatively large without causing a failure to converge or a jump to another branch. We therefore give next a table for larger stepsizes.

For ρ-continuation even very large steps are possible while convergence is lost for μ-continuation.

ρ	μ	x_8	ρ-cont.	μ-cont.	x_8-cont.
6	3.0458	.66391	2	4	(8)
10	3.5010	1.1196	2	6	(9)
14	3.3456	1.5839	2	-	(8)
18	2.9087	2.0554	2	6	
22	2.3837	2.5329	2	5	
26	1.8745	3.0152	2	5	
30	1.4295	3.5014	2	5	
34	1.0643	3.9908	2	5	

Table 2.1 Following the relevant branch for (2.2), (2.3a) with $\varepsilon=0$, $n=15$ starting from $\rho=2$.

ρ	ρ-cont.	μ-cont.
10	3	7
18	3	-
26	3	no conv.
34	3	7
18	3	-
34	3	no conv.
34	4	no conv.

Table 2.2 Continuation with larger steps starting from $\rho = 2$.

3. Following the Spurious Solution Curves

We turn now to the computation of the spurious solution branches of (2.2), (2.3a). Since these branches start at very high values of ρ corresponding to a very small μ if n is large, we have chosen n = 3. We have

$$(3.1) \qquad (B_3^{(1)})^{-1} = \frac{1}{64}\begin{bmatrix} 3 & 2 & 1 \\ 2 & 4 & 2 \\ 1 & 2 & 3 \end{bmatrix} .$$

The relevant solution is asymptotically proportional to the second column of this matrix while there are two branches corresponding to the first (the 1-branch) and the third column of this matrix which, however coalesce in a ρ-μ-diagram. From Theorem 2 in [1] we know that each of these branches "turns back" as a branch which asymptotically is proportional to a linear combination of the first and the second (the γ-branch) respectively the second and the third column. In the first case this linear combination is

$$z^1 = (1,1,1/2)^T. \tag{3.2}$$

Tables 3.1, 3.2 contain the results for following down the 1-branch, it turns back slightly below $\rho = 42$. The starting value was that for $\rho = 74$, $\mu = 3.752E-5$, $x_2 = 1o.754$. An iteration number in parantheses indicates convergence to the relevant branch while a number in double parantheses corresponds to convergence to the γ-branch. Analogous results for the γ-branch in Tables 3.3 and 3.4 show a smaller gain in efficiency for ρ-continuation compared to x_2-continuation but the same robustness.

The performance of the continuation procedure along a component of the solution depends on the choice of this component. The additional work needed for finding an optimal choice would have to be taken into account. We have decided instead to always use the (n+1)/2-th component.

ρ	μ	x_2	ρ-cont.	μ-cont.	x_2-cont.
7o	8.429 E-5	1o.198	2	5	6
66	1.887 E-4	9.6482	2	5	6
62	4.2o5 E-4	9.1o71	2	5	6
58	9.326 E-4	8.5782	2	5	6
54	2.o57 E-3	8.o675	3	5	6
5o	4.5o4 E-3	7.586o	3	5	6
46	9.783 E-3	7.1617	3	7	6
42	2.114 E-2	7.o351	6	8	((5))

Table 3.1 Following the spurious 1-branch for (2.2), (2.3a), $\varepsilon = o$, $n = 3$ starting from $\rho = 74$.

ρ	ρ-cont.	μ-cont.	x_2-cont.
66	2	6	7
58	3	6	7
5o	3	(36)	8
42	6	9	((7))
58	3	8	1o
42	7	((11))	((1o))
42	7	13	((14))

Table 3.2 Results corresponding to Table 3.1 for $\delta\rho$ = 8, 16, 32.

ρ	μ	x_2	ρ-cont.	μ-cont.	x_2-cont.
7o	1.152 E-4	13.6o1	3	5	4
66	2.461 E-4	12.761	3	5	4
62	5.239 E-4	11.916	4	5	4
58	1.111 E-3	11.o62	3	5	4
54	2.345 E-3	1o.196	3	5	3
5o	4.926 E-3	9.3o64	3	5	4
46	1.o28 E-2	8.3685	4	5	6
42	2.121 E-2	7.1444	6	8	((5))

Table 3.3 Following the spurious γ-branch for (2.2), (2.3a), ε = o, n = 3 starting from ρ = 74.

ρ	ρ-cont.	μ-cont.	x_2-cont.
66	3	6	5
58	3	6	5
5o	4	6	((1o))
42	7	9	((6))
58	3	8	((7))
42	7	11	((9))
42	7	no conv.	((13))

Table 3.4 Results corresponding to Table 3.3 for $\delta\rho$ = 8, 16, 32.

In order to show how close the solutions on the different branches are to their asymptotic values we have added Table 3.5.

ρ	branch	x_1	x_2	x_3
	2	5.2781	10.500	5.2781
42	1	9.0292	7.0351	3.5403
	$\tilde{1}$	9.0131	7.1444	3.5964
	2	9.2509	18.500	9.2509
74	1	16.021	10.754	5.3773
	$\tilde{1}$	15.398	14.367	7.2208

Table 3.5 Relevant and spurious solutions for (2.2), (2.3a), $\varepsilon = 0$, $n = 3$ and different values of ρ.

4. Path Following near Turning Points

In 2. we have seen already how the different continuation methods perform when a turning point is present. Here we shall investigate the behaviour in the neighbourhood of a turning point. We have chosen the example (2.2), (2.3a) with $\varepsilon = .2$, $n = 7$. The solution curve has two turning points, for a plot see, for example, [5], and we compute points in the neighbourhood of the second turning point starting from the solution with $\rho = 2$, $\mu = 1.9414$, $x_4 = .31129$. μ-continuation is not possible there, so in Table 4.1 are given the results for ρ- and x_4-continuation. The number of iterations for the latter method is at least twice as large as for algorithm (1.5). Table 4.2 shows that large steps lead only to a very slight increase of this number for ρ-continuation while x_4-continuation may not converge at all.

ρ	μ	x_4	ρ-cont.	x_4-cont.
22	4.4507	3.5349	3	11
42	3.8824	6.7729	2	7
62	3.5936	9.9873	2	5
82	3.4827	13.181	2	5
102	3.4670	16.358	2	4
122	3.5061	19.521	2	4
142	3.5788	22.674	2	4
162	3.6736	25.819	2	4

Table 4.1 Following the relevant branch for (2.2), (2.3a), $\varepsilon = .2$, $n=7$ starting from $\rho = 2$.

ρ	ρ-cont.	x_4-cont.
42	3	no conv.
82	2	8
122	2	5
162	2	5
82	3	no conv.
162	2	12
162	2	no conv.

Table 4.2 Results corresponding to Table 4.1 for $\delta\rho$ = 4o, 8o, 16o.

5. Path Following near Bifurcation Points

In a bifurcation point the matrix on the right side of (1.4) is no longer invertible and the question arises how the algorithm (1.5) performs if used for approaching a bifurcation point. We consider as a simple example the pendulum equation

(5.1)
$$u'' + \mu \sin u = o \quad \text{in } [o,1],$$
$$u(o) = u(1) = o.$$

According to (2.2), (2.3a) we discretize it yielding

(5.2)
$$\mu \begin{bmatrix} \sin x_1 \\ \vdots \\ \sin x_n \end{bmatrix} = B_n^{(1)} x \ .$$

In Table 5.1 we have chosen the same sequence of points which was used in Table 29, [2] p. 195. We start with the solution for μ = 2o, continue with μ-continuation to μ = 1o and switch then to x_5-continuation down to .o2 close to the first (trivial) bifurcation point. The last column gives the number of iterations for algorithm (1.5) using the corresponding ρ-sequence.

μ	ρ	x_5	μ/x_5-cont.	ρ-cont.
15	12.604	1.7613	5	3
14	11.586	1.6259	5	3
13	10.362	1.4602	5	3
12	8.8211	1.2485	5	3
11	6.7111	.95405	6	3
10	2.8880	.41241	7	3
9.8997	2.0999	.3	6	2
9.8378	1.3995	.2	6	2
9.8009	.69964	.1	7	2
9.7965	.55970	.08	5	2
9.7931	.41977	.06	6	2
9.7906	.27984	.04	6	2
9.7891	.13992	.02	7	2

Table 5.1 Path following for (5.2), n = 9 starting from μ = 20.

μ	μ/x_5-cont.	ρ-cont.
13	6	3
10	8	3
9.7965	8	2
9.7891	8	2
10	10	3
9.7891	10	3
9.7891	(6)	3

Table 5.2 Results corresponding to Table 5.1 with larger steps

Again we see that even extremely large continuation steps are possible for (1.5). A number in parantheses indicates convergence to a non-positive solution.

6. How to avoid Spurious Solutions

In the preceding sections we have seen that it was possible to follow the relevant and the spurious solution branches. In all the computations we have performed the ρ-continuation never jumped from a relevant to a spurious path or from a spurious path to one of the other spurious branches. In the following we shall give an explanation for this in the case of the relevant branch.

Although the stepsize $\delta\rho$ may be chosen very large for algorithm (1.5) it is not unbounded as we shall see. In particular if the relevant solution is to be computed it would be desirable to be sure that the iteration leads to a point on that curve and that eventually $\delta\rho$ may be chosen even larger than for (1.5).

We assume now that the matrix $F_o - \lambda_o B$ in Theorem 1.6 is negative-definite on $\{x_o\}^{\perp}$, i. e. x_o is a strict local maximum of f on ∂S_ρ. Then the following modification of algorithm (1.5) is a special case of algorithm (4.3) in [4].

A globally convergent algorithm

(6.1) Let $x_1 \in \partial S_\rho$ be given. For $k = 1,2,\ldots$ do

1. Replace $F_k - \lambda_k B$ in (1.4) by $F_k - \lambda_k B - \tau_k I_n$, where $\tau_k = \max\{o, \delta + \sigma_k\}$ and σ_k is the largest eigenvalue of $F_k - \lambda_k B$ on $\{x_k\}^{\perp}$, $\delta > o$ a given constant. Compute $p_k = -H_k \nabla f(x_k)$.

2. Determine a steplength $\alpha_k = 2^{-j}$ where

$$j = \min\left\{i \in N \cup \{o\}, f\left(\rho \frac{x_k + 2^{-i} p_k}{\|x_k + 2^{-i} p_k\|_B}\right) - f(x_k) \geq 2^{-i-2} |p_k^T \nabla f(x_k)|\right\}$$

and set

$$x_{k+1} = \rho(x_k + \alpha_k p_k) / \|x_k + \alpha_k p_k\|_B .$$

In the first step of (1.6) the matrix $F_k - \lambda_k B$ is regularized since it need not be negative-definite away from the solution curve. Then p_k will be a direction of ascent. In the second a steplength α_k is determined according to the Goldstein-Armijo rule since $\alpha_k = 1$ may not lead to an increase in f. It was proved in [4] that under mild assumptions (6.1) converges globally to a solution of (1.1), that $\alpha_k = 1$, $\tau_k = o$ for $k \geq k_o$ and hence that the order of convergence is two. In general, however, it may be any local maximum to which (6.1) converges. There is obviously a situation when convergence to a certain solution may be guaranteed, namely if one looks for the global maximum x_o of f on ∂S_ρ and $x_1 \in \partial S_\rho$ has a function value which is larger than that of any other critical point. It will now be shown that this simple observation may be utilized for the computation of the relevant solution branches.

In the following we repeatedly refer to [1]. In particular we assume that the problem

$$\nabla f(x) = \lambda B x \tag{6.2a}$$

is such that

$$f(x) = \sum_{i=1}^{n} F(x_i), \quad F : \mathbb{R} \to \mathbb{R}, \tag{6.2b}$$

and hence

$$\nabla f(x) = (F'(x_1), \ldots, F'(x_n))^T.$$

If $g(x) = F'(x) : (o,\infty) \to (o,\infty)$ is continuously differentiable and satifies the growth conditions

$$\frac{1}{g(r)} \sup_{o<\tau\leq t} g(\tau r) \to o, \tag{6.3a}$$

$$\frac{r}{g(r)} \sup_{o<\tau\leq t} g'(\tau r) \to o, \tag{6.3b}$$

as $r \to \infty$ for every $t \in (o,1)$,

then Theorem 1 in [1] states that for every column $b_j = (b_{j1}, \ldots, b_{jn})^T$ of B^{-1} with

(6.4) $$b_{jj} > b_{ij} > o, \quad \forall i \neq j,$$

there is a branch of positive solutions of (6.2)(the j-branch) which is asymptotically proportional to b_j.

We assume now that F and F' satisfy (6.3). Hence, in order to show that for all sufficiently large $\rho > o$ the positive vectors $x,y \in \partial S_1$ satisfy $f(\rho x) > f(\rho y)$ it suffices to show that

(6.5) $$\|x\|_\infty > \|y\|_\infty ,$$

where $\|x\|_\infty = \max\limits_{1 \leq i \leq n} x_i$. It is also sufficient for $\rho > o$ if $F'(t) = t^\alpha, \alpha \geq \alpha_o > o$.

The solution of (6.2) corresponding to b_j in (6.4) attains its maximum in the j-th component for all sufficiently large ρ and for an asymptotic analysis it is sufficient to consider

(6.6) $$x^j = x^j_\rho = \rho b_j / \|b_j\|_B = \rho b_j (b_{jj})^{-1/2}.$$

For simplicity we now assume $B^{-1} > o$.

<u>Lemma 6.7</u> Every vector x^j as in (6.6) for which the column b_j of $B^{-1} > o$, $B \in \mathbb{R}^{n,n}$ symmetric and positive definite, satisfies (6.4), is asymptotically a strict local maximum of f in (6.2) on ∂S_ρ, if F, F' satisfy (6.3a). If (6.4) is valid for $j = 1,\dots,n$ then the strict global maximum is attained for x^j if

$$x^j_j > \max_{i \neq j} x^i_i .$$

<u>Proof</u> The last statement is obvious from the above remarks. It remains to show that for all sufficiently small $|\varepsilon|$

(6.8) $$\|(b_j+\varepsilon v)/\|b_j+\varepsilon v\|_B\|_\infty < b_{jj}^{1/2} ,$$

where $\|v\|_B = 1$. Without restriction of generality we may assume that $v \in \{b_j\}^\perp$. The vector on the left has for all sufficiently small $|\varepsilon|$ its maximum in the j-th component. The left side of (6.8) is then equal to $b_{jj}(b_{jj}+\varepsilon^2)^{-1/2}$ which proves the lemma.

For the special case $F(x)=e^x$ it is a consequence of Theorem 2 in [1] that for any two columns b_j, b_k of B^{-1} which satisfy (6.4), $b_{jj} < b_{kk}$ and

$$b_{ik} \leq b_{jk}, \ \forall i \neq k \text{ or } b_{ji} \leq b_{jk}, \ \forall i \neq j, \tag{6.9}$$

there is a branch (the $\tilde{j}$-branch) of positive solutions of (6.2) which is asymptotically proportional to

$$z^j = \alpha b_j + \beta b_k, \tag{6.10}$$

where $\alpha = (b_{kk} - b_{jk})/\Delta$, $\beta = (b_{jj} - b_{jk})/\Delta$ and $\Delta = b_{jj}b_{kk} - b_{jk}^2 > 0$. This solution attains its maximum in the j-th component which, however, asymptotically in general is not strictly greater than the other components (cf.(3.2)). The question arises if these critical points of f on ∂S_ρ are local extrema.

<u>Lemma 6.11</u> The solutions $\tilde{x}^j = \tilde{x}^j_\rho$ of (6.2) corresponding to z^j in (6.10) satisfy

$$f(\tilde{x}^j) < f(x^j), \ \rho \geq \rho_0, \tag{6.12}$$

x^j as in (6.6). They are in general no local extrema of f on ∂S_ρ.

<u>Proof</u> The inequality (6.12) follows from

$$z^j_j / \|z^j\|_B < b_{jj}^{1/2}$$

which is equivalent to $\Delta > 0$. In order to prove the second assertion we show that for the special choice $B = B_3^{(1)}$ and z as in (3.2) there are vectors $y_1, y_2 \in \mathbb{R}^n$ and numbers $\varepsilon_1, \varepsilon_2 \in \mathbb{R}$ such that

$$\|(z+\varepsilon y_1)/\|z+\varepsilon y_1\|_B\|_\infty > \|z/\|z\|_B\|_\infty, \ 0 < \varepsilon < \varepsilon_1, \tag{6.13a}$$

$$\|(z+\varepsilon y_2)/\|z+\varepsilon y_2\|_B\|_\infty < \|z/\|z\|_B\|_\infty, \ 0 < \varepsilon < \varepsilon_2. \tag{6.13b}$$

Simple calculations show that (6.13a) is valid for $y_1=(0,1,0)^T$ and $\varepsilon_1=4$.

Analogously we find that (6.13b) holds for $y_2=(o,o,1)^T$ and $\varepsilon_2 = 1/2$.

The above lemma has the consequence that with numerical methods which yield only local maxima as, for example, the simple inverse iteration considered in [3] the solutions on the $\tilde{j}$-branches may not be computed. This was in fact observed in numerical tests.

It is obvious how the simple observations in the above lemmas may be utilized, for example, to avoid that the iteration jumps off the relevant solution path. We assume that these solutions correspond to the column of B^{-1} with maximal diagonal element and for all $\rho > o$ strictly maximize f on ∂S_ρ. Following the corresponding path from ρ with solution x_ρ to $\rho + \delta\rho$ the starting solution $(\rho + \delta\rho)/\rho\, x_\rho$ for all sufficiently small $|\delta\rho|$ will have a f-value which is greater than that of any irrelevant solution on level $\rho+\delta\rho$. Then algorithm (6.1) will converge to the relevant solution. On the other hand the results of the above lemmas show that the same situation prevails for sufficiently large ρ, $\delta\rho > o$.

We conclude this section with some numerical results. Computing the relevant solution x_ρ for problem (2.2), (2.3a), $\varepsilon = o$, $n = 3$ the function value of $(\rho+\delta\rho)/\rho\, x_\rho$ is, for example, for $\rho \geq 1$ always considerably larger than that of the solution on the spurious branches. Hence it remains to assure that the algorithm yields an ascent in each step. Continuing upwards from $\rho = 1$, $\mu = 1.4783$ algorithm (1.5) requires 5 iterations for $\delta\rho = 16$, $\rho = 17 \triangleq \mu = .86o16$, and the sequence of function values is increasing. The turning point is at $\rho = 5.o455$, $\mu = 3.3971$. For $\delta\rho = 17$ the algorithm needs 6 iterations but the ascent property is lost, while $\delta\rho = 18$ leads to a negative solution.

Using now only step 2.of (6.1), i.e. the Goldstein-Armijo damping, $\delta\rho$ may be chosen up to 21, $\rho = 22$ corresponds to $\mu = .33741$, and the number of iterations is 3 or 4. For $\delta\rho = 22$ p_k was no direction of ascent and the regularization would have been necessary. Continuing from $\rho = 1o$, $\mu = 2.4392$ with (1.5) works until $\delta\rho = 17$ and needs up to 6 iterations while (6.1) with steplength choice allows $\delta\rho \leq 34$

($\rho=44\hat{=}\mu=2.9E-3$) and requires 3 or 4 iterations. The corresponding figures for $\rho = 12$ are $\delta\rho \leq 22$ for (1.5) and $\delta\rho \leq 52$ for the damped algorithm. $\rho = 64$ corresponds to $\mu = 2.88_{10}-5$. For the values on the irrelevant branches we refer to 3.

7. A Multi-Grid Method for Nonlinear Eigenvalue Problems

In the first six sections we have always chosen discretizations of problems in one space dimension. Most results are very similar if higher dimensional problems are considered. Computations for problem (2.2), (2.3b) and even in three space dimensions showed that algorithms (1.5), (6.1) have equally good numerical properties. The results presented so far have been produced on a microcomputer and should be very easy to reproduce.

If now we turn to discretizations in higher space dimensions then the question of the required computational work becomes even more important. The theoretical properties and the performance of the generalized inverse iteration suggest that it is an ideal algorithm to be further accelerated by, for example, combining it with the multi-grid idea. From the above comparison with continuation procedures using Newton's method one should expect in particular that a proper combination should have better properties than a Newton-based multi-grid algorithm. There are several ways to combine (1.5) with multi-grid ideas. We present next a slight modification of an algorithm for which a convergence proof is given in [5].

We assume that (1.1) is a discretization of a corresponding continuous problem in $N \geq 2$ space dimensions and that we have a sequence of grids $G^{(o)}, G^{(1)}, \ldots, G^{(\ell)}$ with grid constants $h^{(o)} > h^{(1)} > \ldots > h^{(\ell)} > o$ yielding a sequence of discretizations

$$\nabla f^{(i)}(x^{(i)}) - \lambda^{(i)} B^{(i)} x^{(i)} = o, \quad i = o, \ldots, \ell. \tag{7.1}$$

For simplicity we assume that $h^{(i)} = 2h^{(i+1)}$, $i = o, \ldots, \ell-1$. Let I_i^{i+1} (I_{i+1}^{i}) denote interpolation (restriction) operators mapping the functions on $G^{(i)}$ ($G^{(i+1)}$) onto those on $G^{(i+1)}$ ($G^{(i)}$), $i = o, \ldots, \ell-1$. We use further the notation $\|\cdot\|_i = \|\cdot\|_{B^{(i)}}$ and $\rho^{(i)} = \rho^{(o)} 2^{iN/2}$,

$i = o,\dots,\ell$. At first we define a two-grid method on $G^{(o)}$, $G^{(1)}$.

A Two-Grid Method

(7.2) Let $\rho^{(o)} > o$ be given. Set $k = 1$.

1. Compute $x^{(o)}$ with $\|x^{(o)}\|_o = \rho^{(o)}$ and $\lambda^{(o)}$ using algorithm (1.5).

2. Interpolate $x^{(o)}$ to $\tilde{x}_k^{(1)} = I_o^1 x^{(o)}$ and smooth that by performing $\nu^{(1)}$ SOR-steps for the solution of the linear system

$$B^{(1)}x = \nabla f^{(1)}(\tilde{x}_k^{(1)})/\lambda^{(o)} .$$

For the result $x_k^{(1)}$ compute $\lambda_k^{(1)} = \nabla f^{(1)}(x_k^{(1)})^T x_k^{(1)}/\|x_k^{(1)}\|_1^2$.

3. Compute $r_k^{(1)} = -\nabla f^{(1)}(x_k^{(1)}) + \lambda_k^{(1)} B^{(1)} x_k^{(1)}$ and solve

$$\begin{bmatrix} F^{(o)}(x^{(o)})-\lambda^{(o)}B^{(o)} & -B^{(o)}x^{(o)} \\ -x^{(o)T}B^{(o)} & o \end{bmatrix} \begin{bmatrix} \delta^{(o)} \\ \bar{\delta} \end{bmatrix} = \begin{bmatrix} I_1^o\, r_k^{(1)} \\ o \end{bmatrix}$$

4. Set $\tilde{\tilde{x}}_k^{(1)} = x_k^{(1)} + I_o^1\,\delta^{(o)}$, smooth it as in 2. using the system

$$B^{(1)}x = \nabla f^{(1)}(x_k^{(1)})/\lambda_k^{(1)}$$

and normalize the result on level $\rho^{(1)}$ yielding $x_{k+1}^{(1)}$. If a suitable termination criterion is satisfied, stop, otherwise set $k=k+1$ and go to 3.

The only difference to algorithm (3.3) in [5] is the normalization in step 4. This algorithm was devised for computing points on the solution curve for a finer discretization if a corresponding point on the coarser grid is known. If (7.2) is used for continuing along a branch from, say, $\rho_1^{(o)}$ to $\rho_2^{(o)} = \rho_1^{(o)} + \delta\rho^{(o)}$ then it is not exploited in step 2. that the solution $x^{(1)}$ obtained on grid $G^{(1)}$ for $\rho_1^{(1)}$ will be a good starting value for $\rho_2^{(1)}$. Hence step 2. may in this case be replaced by

2'. Set $x_k^{(1)} = x^{(1)}\rho_2^{(1)}/\rho_1^{(1)}$ and compute $\lambda_k^{(1)}$.

The Multi-Grid Method

(7.3) Let $\rho^{(o)} > o$ be given. Set $k = 1$.

1. Compute $x^{(o)}$ with $\|x^{(o)}\|_o = \rho^{(o)}$ and $\lambda^{(o)}$ using algorithm (1.5). With the two-grid method (7.2) compute points $x^{(i)}$, $i=1,\dots,\ell-1$ with $\|x^{(i)}\|_i = \rho^{(i)}$ using the grids $G^{(i-1)}$, $G^{(i)}$ and $\nu^{(i)}$ smoothing steps.

2. Interpolate $x^{(\ell-1)}$ to $\tilde{x}_k^{(\ell)} = I_{\ell-1} x^{(\ell-1)}$ and smooth that by performing $\nu^{(\ell)}$ SOR-steps for the system

$$B^{(\ell)} x = \nabla f^{(\ell)}(\tilde{x}_k^{(\ell)})/\lambda^{\ell-1} .$$

For the result $x_k^{(\ell)}$ compute $\lambda_k^{(\ell)}$.

3. Compute $r_k^{(\ell)} = -\nabla f^{(\ell)}(x_k^{(\ell)}) + \lambda_k^{(\ell)} B^{(\ell)} x_k^{(\ell)}$ and solve

$$\begin{bmatrix} F^{(o)}(x^{(o)}) - \lambda^{(o)} B^{(o)} & -B^{(o)} x^{(o)} \\ -x^{(o)T} B^{(o)} & 0 \end{bmatrix} \begin{bmatrix} \delta^{(o)} \\ \bar{\delta} \end{bmatrix} = \begin{bmatrix} I_1^o \dots I_\ell^{\ell-1} r_k^{(\ell)} \\ 0 \end{bmatrix}$$

4. Set $\approx{x}_k^{(\ell)} = x_k^{(\ell)} + I_{\ell-1}^{\ell} \dots I_o^1 \delta^{(o)}$, smooth it as in 2. using the system

$$B^{(\ell)} x = \nabla f^{(\ell)}(x_k^{(\ell)})/\lambda_k^{(\ell)}$$

and normalize the result on level $\rho^{(\ell)}$ yielding $x_{k+1}^{(\ell)}$. If a suitable termination criterion is satisfied, stop, otherwise set $k = k + 1$ and go to 3.

In (7.3) analogous replacements could be made for path following. Theorem 5.4 in [5] assures local convergence of the above algorithms if only sufficiently many smoothing steps are executed. With algorithms (7.2), (7.3) it was possible to follow the relevant solution branch for problem (2.2),(2.3b), $h^{(o)} = 1/4$ without difficulties up to arbitrarily large ρ-values. The spurious j-branches (cf. 6.) may also be computed while the $\tilde{j}$-branches are no points of attraction of the iteration because of the relationship to simple inverse iteration in the smoothing step.

We give next some results showing that it may be crucial to use step 2'. Problem (2.2), (2.3b) with $h^{(o)} = 1/4$ has essentially two different spurious branches corresponding to the solutions which attain their maximum in the point (●) close to the corner of the unit square or close to the midpoint of an edge.

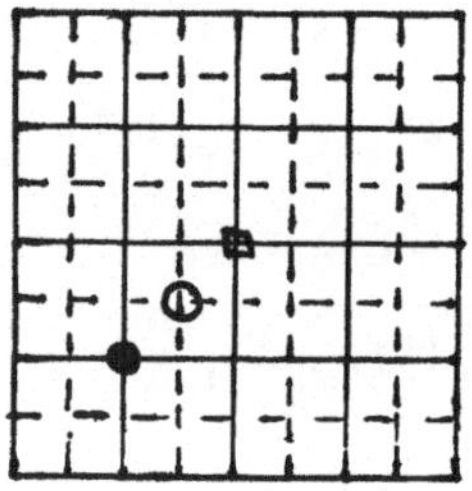

The first of these branches extends down to approximately $\rho^{(o)} = 32.1$ while the corresponding branch for $h^{(1)} = 1/8$ ends somewhat below $\rho^{(1)} = 74$, i. e. $\rho^{(o)} = 37$. If we follow it with algorithm (7.2) from $\rho^{(o)} = 5o$ with $\delta\rho^{(o)} = -1$ then the iteration jumps for $\rho^{(o)} = 45$ to the solution which attains its maximum in the point marked by "o" while for $\rho^{(o)} = 36$ it jumps to the relevant solution ("□"). If now step 2'. is used then the iteration follows the first branch down to $\rho^{(o)} = 37$. The number of two-grid cycles varied between 3 and 7 for $\rho^{(o)} \geq 4o$ and went up below that. With respect to computing time the optimal number of smoothing steps was $\nu^{(1)} = 3$.

References

[1] Beyn, W.-J. and Lorenz, J., Spurious solutions for discrete superlinear boundary value problems. Computing 28, 43-51 (1982)

[2] Bohl, E. Finite Modelle gewöhnlicher Randwertaufgaben. B. G. Teubner Verlag, Stuttgart 1981

[3] Georg, K., On the convergence of an inverse iteration method for nonlinear elliptic eigenvalue problems. Numer. Math. 32, 69-74 (1979)

[4] Mittelmann, H. D., An efficient algorithm for bifurcation problems of variational inequalities. Manuscript NA-81-14, Computer Science Dept., Stanford University 1981 (submitted for publication)

[5] Mittelmann, H. D. Multi-grid methods for simple bifurcation problems. To appear in: W. Hackbusch, U. Trottenberg (eds.), Proceedings of the conference on multi-grid methods, Cologne 1981, Springer-Verlag

[6] Paige, C. C. and Saunders, M. A., Solution of sparse indefinite systems of linear equations. SIAM J. Numer. Anal. 12, 617-629 (1975)

A DEVICE FOR THE ACCELERATION OF CONVERGENCE OF A MONOTONOUSLY ENCLOSING ITERATION METHOD

H. Cornelius and G. Alefeld

Inst. f. Angew. Mathematik
Universität Karlsruhe
Kaiserstrasse 12

7500 Karlsruhe
Germany

1. Introduction

In this paper we consider a class of iteration methods for solving simultaneous systems of nonlinear equations. These methods compute in each iteration step lower and upper bounds for all components of the unknown solution vector. Enclosing the solution repeatedly is under practical consideration advantageous since rounding outwards in a systematic manner one has guaranteed error bounds for the solution. Especially for very large systems this seems to be of great importance since one has observed that - using an arbitrary iteration method - the method comes to a rest although the iterates are still far away from the solution.

The main advantage of the methods considered in this paper consists in the fact that for certain classes of problems (which actually occur in practice) they are convergent to the solution under weaker conditions than known methods which also enclose the solution monotonously. For example, we don't have to assume convexity or similar conditions from which convexity can be derived. If these methods are applied to large systems which originate from the approximation of partial differential equations then the convergence is extremely slow. In this paper we discuss a simple device for constructing a sequence of real vectors which is faster convergent to the solution than the bounds of the enclosing vectors.

2. The method (INSI) and some theoretical results

We assume that the reader has a certain knowledge of interval-analysis to the extent one can find it, for example, in [2].All facts from interval-analysis which are only mentioned without proof can be found in [2]. In this paper we denote real numbers and real n-vectors by $x,y,\ldots$. Real matrices are denoted by $X,Y,\ldots$. Real compact intervals and vectors, the components of which are intervals, are denoted by $[x],[y],\ldots$. Similarly interval-matrices are denoted by $[X],[Y],\ldots$. $d([x])$ denotes the width (or diameter) of the interval $[x]$. $|[x]| = \max_{x\in[x]} |x|$ is called absolute value of the interval $[x]$. For interval-vectors and interval-matrices these concepts are defined via the elements. For example, if $[A] = ([a_{ij}])$ is an n by n interval-matrix then $d([A]) = (d([a_{ij}]))$ is a real n by n matrix. If $f : \mathbb{R}^n \to \mathbb{R}^n$ has a derivative then $f'([x])$ denotes the so-called interval-arithmetic evaluation of the derivative over the interval-vector $[x]$. See, for example [2], Section 3. The interval-arithmetic evaluation of the derivative is an interval-matrix. We consider the splitting

$$f'([x]) = D([x]) - L([x]) - U([x]) \qquad (1)$$

of this matrix where $D([x])$ denotes the diagonal part and where $L([x])$ and $U([x])$ are the parts below and above the main diagonal, respectively.

We start with an interval-vector $[x]^o$ and consider a sequence of interval-vectors $[x]^k$, which are computed by the following iteration method:

$$\left.\begin{array}{l}\left\{\begin{array}{l}\text{Choose } m([x]^k) \in [x]^k \quad (m([x]^k) \text{ a real n-vector})\\ [y]^{k+1} = m([x]^k) - \tilde{D}([x]^k)\,\{L([x]^k)(m([x]^k) - [y]^{k+1}) + \\ \qquad\qquad + U([x]^k)(m([x]^k) - [x]^k) + f(m([x]^k))\}\\ [x]^{k+1} = [y]^{k+1} \cap [x]^k\end{array}\right.\\ k = 0,1,2,\ldots\ .\end{array}\right\} \qquad (2)$$

The diagonal interval-matrix $\tilde{D}([x]^k)$ is defined in the following manner. Let $D([x]^k) = \text{diag}\,(d_{ii}([x]^k))$. Then $\tilde{D}([x]^k) = \text{diag}(1/d_{ii}([x]^k))$. (Please note that the notation $D([x]^k)^{-1}$ would not make sense since for interval-matrices no inverses exist in the ordinary sense). For clarity we stress the fact that the real n-vector $m([x]^k) \in [x]^k$ can be chosen arbitrarily in the interval-vector $[x]^k$.

The method (2) is called interval-arithmetic version of the Newton-single-step method with forming intersections (INSI). The method was introduced in [1] where, however, only the case was considered that $m([x]^k)$ is the center of $[x]^k$.

The following results hold for (INSI). For our considerations the result about the asymptotic convergence factor is of fundamental importance.

Theorem 1. Let $f : \mathfrak{D} \subseteq \mathbb{R}^n \to \mathbb{R}^n$ be a mapping which has a continuous derivative on the open set $\mathfrak{D}$. Assume that f has a zero x^* in $\mathfrak{D}$. Furthermore we assume that for all $[x]^0 \subseteq \mathfrak{D}$ with $x^* \in [x]^0$ the interval-arithmetic evaluation $f'([x]^0)$ of the derivative exists and that $f'([x]^0)$ is split according to (1). Let $0 \notin d_{ii}([x]^0)$, $1 \leq i \leq n$, where $D([x]^0) = \mathrm{diag}\left(d_{ii}([x]^0)\right)$.

a) If $\rho\left\{(|D(x^*)| - |L(x^*)|)^{-1} \cdot |U(x^*)|\right\} < 1$ (ρ denotes the spectral-radius) where

$$f'(x^*) = D(x^*) - L(x^*) - U(x^*)$$

then $\lim_{k\to\infty} [x]^k = x^*$ for all interval-vectors which have sufficiently small width $d([x]^0)$ and for which $x^* \in [x]^0$ holds.

b) For the asymptotic convergence factor of the method (INSI) it holds that

$$R_1\left((\mathrm{INSI}), x^*\right) \leq \rho\left\{(|D(x^*)| - |L(x^*)|)^{-1} \cdot |U(x^*)|\right\} .$$

If $D(x^*) \geq 0$, $L(x^*) \geq 0$, $U(x^*) \geq 0$ then the equality-sign holds in this last inequality.

∎

For the definition of the asymptotic convergence factor of a method which computes interval-vectors see [2], Appendix A. The very long proof of Theorem 1 can be found in [3].

The convergence result $\lim_{k\to\infty} [x]^k = x^*$ from the preceding Theorem is a local one. Under certain assumptions about the interval-arithmetic evaluation $f'([x]^0)$ of the derivative we can get explicit conditions under which the statement $\lim_{k\to\infty} [x]^k = x^*$ holds. These conditions are as follows:

The interval-matrix $[A] = ([a_{ij}])$ has property (K) iff

$$a^1_{ij}\, a^2_{ij} \geq 0 , \quad 1 \leq i,j \leq n ,$$

holds, where $[a_{ij}] = [a^1_{ij}, a^2_{ij}]$.

In order to formulate the next result we need the concept of an M-matrix. A real n by n matrix $A = (a_{ij})$ is called M-matrix iff $a_{ij} \leq 0$, $i \neq j$, and $A^{-1} \geq 0$. See [8], for example.

Theorem 2. The simultaneous system of nonlinear equations is assumed to have a zero x^* in $\mathfrak{D}$. Furthermore we assume that there exists an interval-vector $[x]^0 \subseteq \mathfrak{D}$ with $x^* \in [x]^0$ for which the interval-arithmetic evaluation $f'([x]^o)$ of the derivative exists. Assume that all real matrices from $f'([x]^o)$ are M-matrices. Then the method (INSI) is well-defined and the following hold:

a) $x^* \in [x]^k$, $k \geq o$.

b) $[x]^o \supseteq [x]^1 \supseteq \ldots \supseteq [x]^k \supseteq [x]^{k+1} \supseteq \ldots$

c) $\lim_{k\to\infty} [x]^k = x^*$

d) $R_1\left((\mathrm{INSI}),x^*\right) = \rho\left(\left(D(x^*) - L(x^*)\right)^{-1} U(x^*)\right)$. ∎

The proof may be performed by using the fact that under the assumptions of this Theorem $f'([x]^o)$ has property (K). For details see [3].

3. Application to elliptic difference equations

We are now going to demonstrate that the assumptions of the preceding Theorem can be realized with nonlinear systems which originate from elliptic boundary problems by replacing the derivatives by finite differences. We consider the partial differential equation

$$- F(x,y,u,u_x,u_y,u_{xx},u_{yy}) = o \quad \text{in} \quad R \subset \mathbb{R}^2$$

and the boundary condition

$$u(x,y) = \gamma(x,y) \qquad \text{on} \quad \partial R .$$

R denotes a simply connected bounded region with boundary ∂R.

We assume that F has derivatives with respect to u_{xx} and u_{yy} for which

$$F_{u_{xx}} \geq m > o, \quad F_{u_{yy}} \geq m > o$$

hold. We choose a fixed step-size h in both directions, replace the derivatives by central difference quotients and obtain - after neglecting the discretization error - at the point (x_i,y_j) the equation

$$- g_{ij}\left(x_i,y_j,u_{ij}, \frac{u_{i+1,j} - u_{i-1,j}}{2h}, \frac{u_{i,j+1} - u_{i,j-1}}{2h},\right.$$

$$\left.\frac{u_{i+1,j} - 2u_{ij} + u_{i-1,j}}{h^2}, \frac{u_{i,j+1} - 2u_{ij} + u_{i,j-1}}{h^2}\right) = o .$$

$u_{i,j}$ is an approximation for the unknown function value $u(x_i,y_j)$. Setting $z_k = u_{ij}$, $z = (z_i)$, these equations may be gathered up to $G_1(z) = O$ where the number n of equations is the same as the number

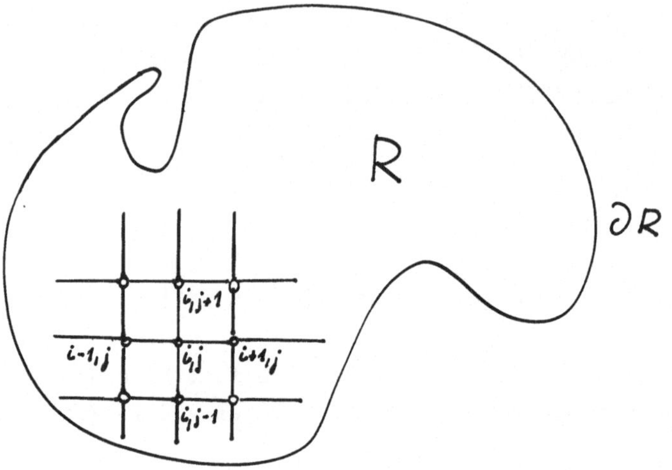

of unknowns. (We omit the details which are necessary to perform the approximation of the differential equation and the boundary conditions for points which are close to the boundary ∂R of R.)

We assume that $F_u \leq 0$ and that h is choosen such small that

$$\left|F_{u_x}\right| \frac{h}{2} < F_{u_{xx}} , \quad \left|F_{u_y}\right| \frac{h}{2} < F_{u_{yy}}$$

hold. Under these conditions it holds that for all $z \in \mathbb{R}^n$ the derivative $G_1'(z)$ is an M-matrix. See [7]. From this it follows that $G_1 : \mathbb{R}^n \to \mathbb{R}^n$ is a so-called M-function, which implies that $G_1(z) = 0$ has at most one solution. See [5] and [6]. Assume now that G_1 is surjective. (Sufficient conditions for the surjectivity of an M-function may be found in [5] and [6]).Then $G_1(z) = 0$ has exactly one solution x^*. By continuity it then follows that for sufficiently small width $d([x]^o)$ of $[x]^o$ and with $x^* \in [x]^o$ all matrices from the interval-arithmetic evaluation $G_1'([x]^o)$ of the derivative of G_1 are M-matrices. Hence the assumptions of Theorem 2 hold for these interval-vectors.

If F has the special form

$$- F(x,y,u,u_x,u_y,u_{xx},u_{yy}) =$$
$$- \left(A(x,y)u_x\right)_x - \left(C(x,y)u_y\right)_y + f(x,y,u)$$

where $A \geq m > 0$, $C \geq m > 0$, $f_u \geq 0$, then it is advantageous to use the following approximations:

$$(Au_x)_x \approx \frac{1}{h^2}\{A(x+\tfrac{h}{2},y)[u(x+h,y) - u(x,y)] - $$
$$- A(x-\tfrac{h}{2},y)[u(x,y) - u(x-h,y)]\},$$

$$(Cu_y)_y \approx \frac{1}{h^2}\{A(x,y+\tfrac{h}{2})[u(x,y+h) - u(x,y)] - $$
$$- A(x,y-\tfrac{h}{2})[u(x,y) - u(x,y-h)]\}.$$

This leads to a nonlinear system of the form

$$G_2(z) = Az + \Phi(z) = 0, \; \Phi(z) = (\phi_i(z_i)),$$

where A is a symmetric, positive definite M-matrix and where the derivative of $\Phi : \mathbb{R}^n \to \mathbb{R}^n$ is isotone. $G_2(z)$ is for arbitrarily choosen step-size h a surjective M-function, that is $G_2(z) = 0$ has exactly one solution x^*.

Furthermore this solution is enclosed by the real n-vectors $z^1 = -A^{-1} \cdot |\Phi(o)|$ and $z^2 = -z^1$:

$$z^1 \leq x^* \leq z^2 .$$

See [4], p. 460. Since Φ is isotone we can set the lower bounds of the diagonal elements of the interval-arithmetic evaluation $\Phi'([x]^o)$ equal to zero if these bounds are negative. Since A is an M-matrix it then follows that all point-matrices which are contained in $G_2'([x]^o)$ are M-matrices, that is the assumptions of Theorem 2 hold. We stress the fact that for the system $G_2(z) = 0$ no assumptions about the width of $[x]^o$ are needed in order that the assumptions of Theorem 2 hold. The only really important assumption about $[x]^o$ is the inclusion $x^* \in [x]^o$ of the solution x^*. As shown above such an $[x]^o$ can be computed by solving a linear system of simultaneous equations.

If one applies (INSI) to systems of the form $G_1(z) = 0$ or $G_2(z) = 0$ then one observes that the convergence is extremely show. This is especially the case if the number of unknowns becomes larger and larger. The reason is that the spectral-radius $\rho\left((D(x^*)-L(x^*))^{-1} U(x^*)\right)$ approaches one if h goes to zero. By part d) of Theorem 2 this implies that the same is the case for the asymptotic convergence factor $R_1((\text{INSI}),x^*)$. For this reason (INSI) can not be considered to be a realistic method for computing the solution of $G_1(z) = 0$ or $G_2(z) = 0$.

On the other hand we consider the fact that (INSI) applied to $G_2(z)$ is convergent to x^* (without important additional assumptions) for all starting interval-vectors which enclose x^* as an advantageous property which we would not like to give up.

4. The proposed modification of (INSI)

In order to compute sequences of vectors which are converging faster to the solution than the bounds of the interval-vectors computed using (INSI) we now use the fact that $m([x]^k) \in [x]^k$ can be choosen arbitrarily in each step of (INSI). Therefore we introduce an instruction for choosing $m([x]^k)$, which uses only data which are already known from (INSI) and for which there is a chance that the sequence $\{m([x]^k)\}$ is faster convergent to x^* than the bounds of $\{[x]^k\}$.

In order to formulate this instruction we need a so-called cut-off function κ which is defined in the following manner:
Let $w \in \mathbb{R}$ and the interval $[x] = [x_1, x_2]$ be given. Then

$$\kappa(w,[x]) = \begin{Bmatrix} x_1 & , & w < x_1 \\ w & , & w \in [x] \\ x_2 & , & w > x_2 \end{Bmatrix} .$$

For $[x] = ([x]_i)$, $u = (u_i)$ we define

$$p(u,[x]) = \big(\kappa(u_i,[x]_i)\big) .$$

Using p we consider the following method, called (INSI) + (SOR), which differs from (INSI) only by adding an explicit rule for the selection of $m([x]^k)$.

$$\left.\begin{array}{l} \text{Choose } m([x]^0) \text{ to be center of } [x]^0. \\ \omega_{-1} := 1. \\ \left\{\begin{array}{l} [y]^{k+1} = m([x]^k) - \tilde{D}([x]^k)\{L([x]^k)\big(m([x]^k)-[y]^{k+1}\big) + \\ \qquad + U([x]^k)\big(m([x]^k)-[x]^k\big)+f\big(m([x]^k)\big)\} \\ [x]^{k+1} = [y]^{k+1} \cap [x]^k \\ \gamma_k = \dfrac{\|d([x]^{k+1})\|_\infty}{\|d([x]^k)\|_\infty} \quad (\text{if } d([x]^k) \neq 0) \\ \omega_k = \begin{cases} \dfrac{2}{1+\sqrt{1-\gamma_k}} & \text{if } \gamma_k \neq 1 \\ \omega_{k-1} & \text{otherwise} \end{cases} \\ \vdots \end{array}\right. \end{array}\right\} \quad (3)$$

$$\left\{\begin{array}{l} u^{k+1} = m([x]^k)-\omega_k \left\{m(D([x]^k))-\omega_k\ m(L([x]^k))\right\}^{-1}\cdot f(m[x]^k) \\ m([x]^{k+1}) = p(u^{k+1},\ [x]^{k+1}) \end{array}\right.$$

$k = 0,1,2,\ldots$.

$\{m(D([x]^k))$ and $m(L([x]^k))$ are arbitrary real matrices which are taken from $D([x]^k)$ and $L([x]^k)$ respectively. Naturally one can choose the centers of these matrices$\}$. In [3],p. 38 ff,it is demonstrated in detail why with the given choice of $m([x]^k)$ one can be rather sure, that the sequence $\{m([x]^k)\}$ converges considerably faster to x^* than the bounds of the sequence $\{[x]^k\}$.

We finally remark that the instruction which is used for computing u^{k+1} (and therefore also for computing $m([x]^{k+1})$) may be considered to be an approximate step of the Newton-SOR-method applied to $f(x) = 0$. (Concerning the Newton-SOR-method see [4], p.217 ff).

In passing we note that instead of performing one approximate step of the Newton-SOR-method one can do the same using any other iteration method which promises to converge faster to x^* than the bounds of the sequence $\{[x]^k\}$. In [3] the use of the ADI-method was discussed in some detail. The numerical results are even much better than with the results listet subsequently for (INSI) +(SOR). However, no theoretical foundation can be given in this case because of the fact that certain matrices do not commute.

5. Numerical examples

Example 1.

As a first example we consider the equation

$$\Delta u = \frac{u^3}{1+x^2+y^2} \quad \text{in} \quad (0,1) \times (0,1)$$

with the boundary conditions

$$u(x,o) = 1 \quad \text{and} \quad u(x,1) = 2-e^x \quad \text{for} \quad x \in [0,1]$$

$$u(o,y) = 1 \quad \text{and} \quad u(1,y) = 2-e^y \quad \text{for} \quad y \in [0,1] .$$

Example 2.

$$\Delta u = e^u \quad \text{in} \quad (0,1) \times (0,1) = R$$

$$u(x,y) = x + 2y \quad \text{on} \quad \partial R$$

(Please note that the results of this paper are not limited to rectangular regions. Numerical examples for which the boundary is curvilinear are given in [3]).

In the tables given subsequently we have compared our results with those from the paper "Aspects of Nonlinear Block-Successive Overrelaxation" by L.A.Hagemann and T.A.Porsching (SIAM J. Numer.Anal., 12, 316-335 (1975)). In that paper a very lengthy instruction is given which forces the normally only local convergent nonlinear block-successive overrelaxation method to converge to the solution x^*. Therefore this modification is comparable to our method (INSI) + (SOR) where by (INSI) convergence is guaranteed for all interval-vectors which enclose the solution.

In both examples the following termination criteria were used:

$$\|d([x]^k)\|_\infty \leq 2\cdot 10^{-6} \quad \text{for} \quad \text{(INSI)}$$

$$\|u^{k+1} - m([x]^k)\|_\infty \leq 10^{-6} \quad \text{for} \quad \text{(INSI) + (SOR)}$$

$$\|x^{k+1} - x^k\|_\infty \leq 10^{-6} \quad \text{for} \quad \text{(H-P)}.$$

In order to make a fair judgement on the proposed method (INSI) + (SOR) one has to take into account that the interval-operations necessary for performing this method have been programmed using subroutines. If there would be available a realization of the interval-operations which - concerning the execution time - is comparable to the usual floating-point operations, then the proposed method would compare even more favorable.

The examples have been computed using a CYBER 175 of the Wissenschaftliches Rechenzentrum Berlin (WRB).

Example 1

	h	$\frac{1}{4}$	$\frac{1}{8}$	$\frac{1}{16}$	$\frac{1}{20}$	$\frac{1}{32}$	$\frac{1}{64}$	$\frac{1}{91}$	$\frac{1}{128}$
	$n=(\frac{1}{h}-1)^2$	9	48	225	361	961	3969		16129
(INSI)	Steps	21	90	366	572	1466	5856*	---	23424*
	Time(sec)	0.067	1.624	30.448	76.368	521.922	2.39* hours	---	38.85* hours
(INSI)+	Steps	11	22	47	61	105	248	400	---
(SOR)	Time(sec)	0.041	0.473	4.605	9.548	45.3	431.8	1396	1.17* hours
(H-P)**	Steps	22	39	69	88	123	237	339	
	Time(sec)	0.04	0.403	2.950	6.081	24.5	183.9	518	

Both for (INSI) and for (INSI) + (SOR) all components of the starting vector $[x]^o$ have been chosen to be the interval [-1,2]. For (H-P) we chose $x^o = (x_i^o)$, $x_i^o = 2$ for all i.

*Estimated values.

**Means the point overrelaxation method using the strategy given by Hagemann and Porsching in the paper mentioned in the text.

Example 2

	h	$\frac{1}{4}$	$\frac{1}{8}$	$\frac{1}{16}$	$\frac{1}{20}$	$\frac{1}{32}$	$\frac{1}{64}$	$\frac{1}{91}$	$\frac{1}{128}$
	$n=(\frac{1}{h}-1)^2$	9	49	225	361	961	3969		16129
(INSI)	Steps	19	81	324	507	1298	-	-	-
	Time(sec)	0.043	0.943	17.54	44.2	301	-	-	-
(INSI)+	Steps	10	21	46	59	102	248	393	-
(SOR)	Time(sec)	0.029	0.311	3.164	6.554	30.4	298.128	987	49*min
(H-P)**	Steps	21	39	77	74	122	251	342	
	Time(sec)	0.051	0.507	4.718	7.911	33.09	258.1	692.5	-

Both for (INSI) and for (INSI) + (SOR) all components of the starting vector $[x]^0$ have been chosen to be the interval [0,3]. For (H-P) we chose $x^0 = (x_i^0)$, $x_i^0 = 3$.

*Estimated value.

**Means the point overrelaxation method using the strategy given by Hagemann and Porsching in the paper mentioned in the text.

References

[1] G.Alefeld : Über die Existenz einer eindeutigen Lösung bei einer Klasse nichtlinearer Gleichungssysteme und deren Berechnung mit Iterationsverfahren.
Aplikace Matematiky 17, 267-294 (1972).

[2] G.Alefeld, J.Herzberger : Einführung in die Intervallrechnung. Bibliographisches Institut, Reihe Informatik 12, Mannheim 1974.

[3] H.Cornelius : Untersuchungen zu einem intervallarithmetischen Iterationsverfahren mit Anwendungen auf eine Klasse nichtlinearer Gleichungssysteme.
Dissertation. Fachbereich Mathematik der TU Berlin, Berlin 1981.

[4] J.M.Ortega, W.C.Rheinboldt : Iterative Solution of Nonlinear Equations in Several Variables.
Academic Press, New York - London 1970.

[5] W.C.Rheinboldt : On M-Functions and their Application to Nonlinear Gauss-Seidel Iterations and Network Flows. Ges.f.Math.u.Datenverarbeitung, Birlinghoven/Germany. Tech.Rep. BMwF-GMD-23. (1969).

[6] W.C.Rheinboldt : On classes of n-dimensional nonlinear mappings generalizing several types of matrices. Numerical Solution of Partial Differential Equations - II. Synspade 1970. B.Hubbard (Ed.).Academic Press, New York - London 1971.

[7] W.Törnig : Monoton einschließend konvergente Iterationsprozesse vom Gauss-Seidel Typ zur Lösung nichtlinearer Gleichungssysteme im $\mathbb{R}^N$ und Anwendungen. Technische Hochschule Darmstadt. Preprint-Nr. 517, Dezember 1979.

[8] R.S.Varga : Matrix Iterative Analysis. Prentice-Hall Inc. Englewood Cliffs, N.J. 1962.

OVERRELAXATION IN MONOTONICALLY CONVERGENT ITERATION METHODS

Bernhard Kaspar
Technische Hochschule Darmstadt
6100 Darmstadt

1 INTRODUCTION

If you are only interested in rapid convergence , relaxation methods cannot compete with modern methods (e.g. multigrid algorithms or (preconditioned) c.g.) . Nevertheless there are some advantages , which we don't want to discuss here (see Gipser [2]) .

A particular role is played by the variant to be considered here , namely methods which produce monotonically converging sequences with respect to both sides . One automatically obtains lower and upper bounds for the solution and by this a reliable termination criterion (which is sometimes a crucial point in other algorithms) . Methods of this type have been investigated (besides others) by Ortega and Rheinboldt [3] , [4] , Schelin [6] and Törnig [8] . For different approaches see also Albrecht [1] and Schröder [7] .

A common fact in these investigations is the restriction to underrelaxation (in the nonlinear case) . In the following we want to show , that with slight modifications one can prove monotone convergence even with overrelaxation . The upper bound for the relaxation parameter can either be given in advance or be computed in an adaptive way .

The theorem on monotone convergence shall only be shown for the special case of SOR - Newton - Type methods . This is done in section 2 . Also generalizations or modifications are outlined , but not proved . In the last section we demonstrate the application by a simple example , namely the system arising in the discretization of a semilinear boundary value problem . Also we report on some (first) numerical experiences .

In the following , some notations are needed .

For $x , y \in R^n$ $\quad x \le y$ means $x_i \le y_i \quad i = 1 (1) n$

and $\langle x , y \rangle$ denotes the corresponding order interval , that is $z \in \langle x , y \rangle$ iff $x \le z \le y$

e_i is the i-th coordinate vector of R^n

$(x / z_i) := (x_1 , \quad , x_{i-1} , z_i , x_{i+1} , \quad , x_n)$

For $F : R^n \to R^n$ $\quad (F = (F_1 , \ldots , F_n)^T) : \quad d_j F_k := \frac{d}{dx_j} F_k$

denotes the elements of the Jacobian F'

2 A CONVERGENCE THEOREM

We consider the following problem

$$\text{Given} \quad F : D \subset R^n \longrightarrow R^n$$
$$\text{find} \quad z^* \in D \quad : \quad F(z^*) = 0 \tag{1}$$

We suppose to have an initial estimate

$$z^* \in \langle x^o , y^o \rangle \qquad \text{where} \qquad F(x^o) \leq F(y^o) \tag{2}$$

We want to investigate the following method :

Given a sequence $i(k)$, $i \in \{1, \quad ,n\}$, $k \in N$

$$\text{set} \quad y^{k+1} := y^k - \bar{\omega}^k \; \bar{t}^k \; F_{i(k)}(y^k) \; e_{i(k)}$$
$$x^{k+1} := x^k - \underline{\omega}^k \; \underline{t}^k \; F_{i(k)}(x^k) \; e_{i(k)} \tag{3}$$
$$\text{where} \quad \bar{t}^k , \underline{t}^k > 0$$

DEFINITION The method (3) is said to be monotonically convergent with respect to both sides

$$\text{iff} \quad y^k \downarrow y^* \quad ; \quad x^k \uparrow x^* \qquad \text{as } k \longrightarrow \infty \tag{4a}$$
$$F(y^{l(k)}) \geq 0 \geq F(x^{l(k)}) \qquad k \in N \tag{4b}$$
$$F(x^*) = 0 = F(y^*) \tag{4c}$$

where $l(k)$ runs over a subsequence of N

REMARK In the usual Definition $l(k) = k$. This condition shall be weakened here .

Notice , that with this Definition $x^* , y^* \in \langle x^k , y^k \rangle$ for all k , so one obtains a reliable termination criterion for the algorithm (3) .

In the following we want to confine ourselvelves to SOR-Newton-Type methods , i.e.

$$\bar{t}^k := (d_{i(k)} F_{i(k)}(y^k))^{-1} \quad ; \quad \underline{t}^k := (d_{i(k)} F_{i(k)}(x^k/y^k_{i(k)}))^{-1}$$

In addition we want to assume the Jacobian $F'(z)$ to be an L - Matrix for $z \in \langle x^o , y^o \rangle$.

To illustrate the algorithm assume

$$F(y^k) \geq 0 \qquad (\text{ so } y^{k+1} \leq y^k)$$

Now for $j \neq i(k)$:

$$F_j(y^{k+1}) = F_j(y^k) + d_{i(k)}F_j(z) \, (y^{k+1} - y^k) \geq F_j(y^k) \geq 0 \qquad (5)$$

$$\text{where} \quad z \in \langle y^{k+1} , y^k \rangle \quad .$$

Also , by a simple Taylor expansion , one gets

$$F_{i(k)}(y^{k+1}) = c^k \, F_{i(k)}(y^k) \qquad (6)$$

$$\text{where} \quad c^k = 1 - \bar{\omega}^k \frac{d_{i(k)}F_{i(k)}(z)}{d_{i(k)}F_{i(k)}(y^k)} \qquad \text{and} \quad z \in \langle y^{k+1} , y^k \rangle \quad .$$

So under suitable assumptions concerning F' , underrelaxation (i.e. $\bar{\omega}^k \leq 1$ or $c^k \geq 0$) yields

$$F_{i(k)}(y^{k+1}) \geq 0$$

and , together with (5) :

$$F(y^{k+1}) \geq 0 \quad .$$

Now we want to allow the relaxation parameter to be $\bar{\omega}^k = 1 + b^k$, where $b^k \geq 0$ and skip the condition

$$F_{i(k)}(y^{k+1}) \geq 0$$

if we can guarantee :

$$F_{i(k)}(y^{k+2}) \overset{!}{\geq} 0$$

i.e. what was done wrong in step k is corrected in step $k+1$. This guarantee must be given even with $\bar{\omega}^{k+1} = 1$, to avoid the relaxation parameters to grow without control . Then of course also with $\omega^{k+1} > 1$ we are on the right side , see (5) .

This results in an upper bound for ω^k , which can be expressed in terms of elements of the Jacobian .

Let us introduce the following numbers

$$c := \inf_{\substack{z,w \in \langle x^o, y^o \rangle \\ k \in K_1}} \left| \frac{d_{j(k+1)}F_{j(k)}(w)}{d_{j(k+1)}F_{j(k+1)}(z)} \right| \quad ; \quad \tilde{c} := \inf_{\substack{z,w \in \langle x^o, y^o \rangle \\ k \in K_2}} \left| \frac{d_{j(k)}F_{j(k+1)}(z)}{d_{j(k)}F_{j(k)}(w)} \right|$$

where

$K_1 := \{k \in N : \ d_{j(k+1)} F_{j(k)}(z) \neq 0\}$ and $K_2 := \{k \in N : \ d_{j(k)} F_{j(k+1)}(z) \neq 0\}$

For abbreviation

$$\bar{a}^k := \frac{F_{i(k+1)}(y^k)}{F_{i(k)}(y^k)} \quad ; \quad \underline{a}^k := \frac{F_{i(k+1)}(x^k)}{F_{i(k)}(x^k)} \qquad (8)$$

We can now state

THEOREM Let

i) $F \in C^2(\langle x^o, y^o \rangle)$

ii) $x^o \leq y^o$ and $F(x^o) \leq 0 \leq F(y^o)$

iii) $F'(z)$ strictly diagonally dominant , $F'(z)$ L - matrix for $z \in \langle x^o, y^o \rangle$

iv) DF' ($:= (d_i F_i)_{i=1(1)n}$) diagonally isotone (e.g. $d_{ii} F_i \geq 0$)

v) $i(k)$ cyclic (i.e. $i(k) = i(k+n)$)

vi) $1 \leq \bar{\omega}^k \leq \dfrac{1 + c\, \bar{a}^k}{1 - c\, \tilde{c}}$; $1 \leq \underline{\omega}^k \leq \dfrac{1 + c\, \underline{a}^k}{1 - c\, \tilde{c}}$

where $\omega^k = 1$ if $d_{i(k+1)} F_{i(k)} = 0$ and $\omega^{l(k)} = 1$ for a subsequence

Then (1) admits a unique solution $z^* \in \langle x^o, y^o \rangle$ and the SOR-Newton iteration (3) is monotonically convergent

Proof Let $F(y^k) \geq 0$. Then $y^{k+1} \leq y^k$ and , for $j \neq i(k)$ (see (5) :

$$F_j(y^{k+1}) \geq 0 \ .$$

Again by Taylor expansion one obtains

$$F_{i(k)}(y^{k+2}) = F_{i(k)}(y^k) -$$

$$\bar{\omega}^{k+1} \left(F_{i(k+1)}(y^k) - d_{i(k)} F_{i(k+1)}(z^3)\, \bar{\omega}^k \frac{F_{i(k)}(y^k)}{d_{i(k)} F_{i(k)}(y^k)} \right) \frac{d_{i(k+1)} F_{i(k)}(z^1)}{d_{i(k+1)} F_{i(k+1)}(y^{k+1})}$$

$$- \bar{\omega}^k F_{i(k)}(y^k) \ \frac{d_{i(k)} F_{i(k)}(z^2)}{d_{i(k)} F_{i(k)}(y^k)}$$

where $z^1, z^3 \in \langle y^{k+1}, y^k \rangle$ and $z^2 \in \langle y^{k+2}, y^{k+1} \rangle$

Hence (because of iv) and $\bar{\omega}^{k+1} \geq 1$) :

$$F_{i(k)}(y^{k+2}) \geq (1 - \bar{\omega}^{k} F_{i(k)}(y^{k})) -$$

$$(F_{i(k+1)}(y^{k}) - d_{i(k)}F_{i(k+1)}(z^{3})\, \bar{\omega}^{k} \frac{F_{i(k)}(y^{k})}{d_{i(k)}F_{i(k)}(y^{k})}) \frac{d_{i(k+1)}F_{i(k)}(z^{1})}{d_{i(k+1)}F_{i(k+1)}(y^{k+1})}$$

$$\overset{!}{\geq} 0$$

Hence

$$\bar{\omega}^{k} \overset{!}{\leq} \frac{1 - \dfrac{d_{i(k+1)}F_{i(k)}(z^{1})}{d_{i(k+1)}F_{i(k+1)}(y^{k+1})} \dfrac{F_{i(k+1)}(y^{k})}{F_{i(k)}(y^{k})}}{1 - \dfrac{d_{i(k+1)}F_{i(k)}(z^{1})}{d_{i(k+1)}F_{i(k+1)}(y^{k+1})} \dfrac{d_{i(k)}F_{i(k+1)}(z^{3})}{d_{i(k)}F_{i(k)}(y^{k})}}$$

This is fulfilled , if

$$\bar{\omega}^{k} \leq \frac{1 + c\, \bar{a}^{k}}{1 - c\, \tilde{c}} \qquad \text{for} \quad d_{i(k+1)}F_{i(k)} \neq 0$$

Analogously

$$\underline{\omega}^{k} \leq \frac{1 + c\, \underline{a}^{k}}{1 - c\, \tilde{c}}$$

Now let l be any number with $\bar{\omega}_{l} = \underline{\omega}_{l} = 1$

Then $$F(y^{l+1}) = 0 = F(x^{l+1})$$

so $$0 = F(y^{l+1}) - F(x^{l+1}) = B\ (y^{l+1} - x^{l+1})$$

where B is the Jacobian evaluated at points z^{j} with different arguments z^{j} in different rows and $z^{j} \in \langle x^{l+1}, y^{l+1} \rangle$ for each j . B is a strictly diagonally dominant L - matrix , hence an M - matrix . Since i(k) is cyclic , the sign of $F_{i(j)}(y^{j})$ or $F_{i(j)}(x^{j})$ respectively , occuring in the algorithm is correct for each j . It follows :

$$y^{l+1} \geq x^{l+1}$$

We now have obtained two monotone and bounded sequences , so

$$y^{k} \downarrow y^{*} \ ; \quad x^{k} \uparrow x^{*}$$

and by an argument of continuity :

$$F(x^*) = 0 = F(y^*) \quad .$$

With assumption iii) of the theorem $x^* = y^*$,
and the proof is complete .

REMARKS

1) Condition iii) could be altered , requiring $F'(z)$ to be irreducible diagonally dominant . Presumption iv) of the theorem is a convexity condition , which causes the different evaluation of $d_i F_i$ in the calculation of x^{k+1} . Of course one could prove an analogous version in the concave case .
The theorem can be extended for more general sequences $i(k)$ (e.g. SSOR - Newton is possible) .

2) The upper bound for ω^k gives a hint , how the sequence $i(k)$ should be choosen .(The parameters c and $\tilde{c}$ should be as large as possible , to accelerate the convergence .) In section 3 we give an example , to show , how these numbers can be computed in a concrete problem .
If you ignore the parameter a^k , you can keep the upper bound for ω^k fixed during the algorithm . Otherwise the ω^k is computed for each step , evaluating in addition $F_{i(k+1)}$. Numerical experiments show , that the additional cost per step can be more then counterbalanced by the decrease of the number of iterations .

3) The theorem can be generalized for algorithms using formula (3) . (For example you could choose $(t^k)^{-1}$ as an approximate of $d_{i(k)}F_{i(k)}$, or keep t^k fixed) .
Instead of $F_{i(k)}(y^{k+2}) \geq 0$ you could postulate $F_{i(k)}(y^{k+p}) \geq 0$, where $2 \leq p < n$ .The upper bound for ω^k then increases , but nevertheless this doesn't seem to be worthwhile , because one needs more evaluations of F .

3 AN EXAMPLE AND NUMERICAL EXPERIMENTS

We want to apply the algorithm to the discretization of a simple semilinear partial differential equation .

$$L\,u = -\Delta u + f(u,x,y) \qquad \text{in a rectangular domain (for simplicity)}$$

with Dirichlet - data on the boundary .

Under suitable assumptions (e.g. $f_u \geq 0$) the problem is of monotonic type (i.e. a maximum principle holds) .
Here you can construct upper and lower solutions v , w , accomplishing

$$L\,v \leq 0 \leq L\,w \qquad \text{and hence} \qquad v \leq u \leq w$$

(c.f. Collatz [3])

The problem is discretized in the most common way , resulting in a system

$$F\,U = 0 \quad \text{where } U \in R^n \ .$$

If the unknowns U_i are ordered lexicographically , the Jacobian F' is then

$$F' = \begin{bmatrix} A & -I & & 0 \\ -I & A & -I & \\ & \ddots & \ddots & \ddots \\ 0 & & \ddots & -I \\ & & -I & A \end{bmatrix} \quad \text{and} \quad A = \begin{bmatrix} 4+0(h^2) & -1 & & 0 \\ -1 & 4+0(h^2) & \ddots & \\ & \ddots & \ddots & \ddots \\ 0 & & \ddots & -1 \\ & & -1 & 4+0(h^2) \end{bmatrix}$$

Notice :
F' is an L - matrix (and therefore , if the discretization parameter h is small enough , an M - matrix) ; $d_{ii}F_i = h^2 f_{u_i} \geq 0$.

We choose the natural lexicographical sequence $i(k)$.

Then $c = \tilde{c} = 1/(4 + h^2 f_{max}) = 1/4 - 0(h^2)$

and $d_{i(k+1)}F_{i(k)} = 0$, whenever the algorithm has passed through a line of the grid .

The initial estimates x^0 , y^0 can be computed as discrete analoga of v and w respectively , so all conditions of the theorem are fulfilled .

The overrelaxation method we proposed has been tested for the above problem (with different choices of f and different mesh sizes) , as well as for a simple quasilinear equation .
With $\omega^k = 1/(1 - c\,\tilde{c})$ fixed we could reduce the number of iterations , to obtain a given accuracy , up to about 85 % , without any further work . If the relaxation parameter was computed adaptively , the number of iterations was about a half , compared with the number obtained with $\omega^k = 1$ fixed . Notice , that in this case we had to carry out 6 evaluations of F or dF respectively per step (instead of 4 , when ω^k is fixed) , so the work to be performed was about 3/4 of the work in the Gauß - Seidel case .
For a corresponding ordinary differential equation , the acceleration of convergence obtained by overrelaxation is even more effective , since here

$$c = \tilde{c} = 1/2 - O(h^2) .$$

Though we need more experiments , to give sufficient information on the effect of overrelaxation , this method seems to give some improvements .

REFERENCES

1 Albrecht , J. " Fehlerschranken und Konvergenzbeschleunigung bei einer monotonen oder alternierenden Iterationsfolge " Numerische Mathematik 4 ,1962

2 Gipser , M. " On global convergence of coordinate relaxation in the case of an unsymmetric , diagonally dominant Jacobian " To appear in Computing

3 Collatz , L. " The numerical treatment of differential equations " Springer Verlag , Berlin , Göttingen , Heidelberg 1960

4 Ortega , J.M. ; Rheinboldt , W.C. " Monotone iterations for nonlinear equations with application to Gauß - Seidel methods " SIAM J. Num. Anal. 4 1967

5 Ortega , J.M. ; Rheinboldt , W.C. " Iterative solution of nonlinear equations " Academic Press New York , London 1970

6 Schelin , W. " Monotone convergence of the SOR-Newton iterative technique " SIAM J. Num. Anal. 32 1973

7 Schröder , J. " Anwendung von Fixpunktsätzen bei der numerischen Behandlung nichtlinearer Gleichungen " Arch. Rat. Mech. Anal. 4 1959

8 Törnig , W. " Monoton einschließend konvergente Iterationsprozesse vom Gauß-Seidel-Typ zur Lösung nichtlinearer Gleichungssysteme im R^n " Math. Meth. in the Appl. Sci. 2 1980

SIMPLE BOUNDS FOR ZEROS OF SYSTEMS OF EQUATIONS

Arnold Neumaier
Institut für Angewandte Mathematik
Universität Freiburg
D-7800 Freiburg, W-Germany

Abstract. Let $F:D(\subseteq \mathbb{R}^n) \to \mathbb{R}^n$ be a continuous function, and suppose that for some $x_0 \in D$ and some nonsingular matrix A the vector $\delta_0 := A^{-1} F(x_0)$ is "small". Assuming the existence of a nonnegative vector $c \in \mathbb{R}^n$ such that

$$|F(x) - F(x_0) - A(x - x_0)| \leq ||\delta_0|| c$$

for all x in a suitable neighbourhood S of x_0, a simple condition is given which guarantees that S contains a zero $\hat{x}$ of F. The resulting bounds are shown to be quite accurate. They contain as a special case the bounds obtainable from a theorem of Kantorovic.

It is discussed how to compute the required vector c, and how to make efficient use of sparsity. The practical use of the bounds is demonstrated by an extensive example, the finite difference equations obtained by discretizing the minimal surface equation.

1. Introduction

Let D be a subset of $\mathbb{R}^n$, and let $F: D \to \mathbb{R}^n$ be a bounded continuous function. We assume the following:

(i) A good approximation $x_0 \in D$ for an isolated zero $\hat{x}$ of $F(x)$ is known.

(ii) An approximation A for the Fréchet-derivative or a generalized slope operator near x_0 is known.

(iii) A is nonsingular, and the structure of A allows the practical solution of linear equations with coefficient matrix A by Gauss elimination.

With these assumptions, we ask for a bound on the magnitude of the error $\hat{x}-x_0$. The bound shall be such that we can take advantage of any special structure of A such as diagonal dominance or sparseness.

Assuming (i), (ii), and (iii), we have for x near x_0

$$F(x) - F(x_0) \approx A(x - x_0), \tag{1}$$

hence, for $x = \hat{x}$ with $F(\hat{x}) = 0$,

$$\hat{x} - x_0 \approx -A^{-1} F(x_0).$$

Thus, if we put

$$\delta_0 := A^{-1} F(x_0), \tag{2}$$

we have in any norm $\|\hat{x}-x_0\| \approx \|\delta_0\|$, and we can expect an error estimate

$$\|\hat{x} - x_0\| \le \kappa \|\delta_0\|$$

with a reasonably small factor $\kappa > 1$. To guarantee $\hat{x} \in D$ we assume that

$$S := \{x \in \mathbb{R}^n \mid \|x - x_0\| \le \kappa \|\delta_0\|\} \subseteq D, \tag{3}$$

and to make (1) precise we require that

$$|F(x) - F(x_0) - A(x - x_0)| \le \|\delta_0\| c \quad \text{for all } x \in S. \tag{4}$$

Hence c is a nonnegative vector; absolute value and order is understood to be componentwise.

It turns out that (2), (3), and (4), together with a simple inequality are sufficient to show that S must contain at least one zero of F(x). The precise statement and its proof is contained in Section 2. Section 3 discusses the practical computation of the quantities involved in the bound. In Section 4 we compare our bounds with those of the Kantorovic theorem, and in Section 5 we present some numerical examples.

The concepts used in this paper can be found, e.g. in the book by Ortega and Rheinboldt [12].

2. The main result

For the moment we shall forget about the heuristic introduction, and suppose that D is a subset of $\mathbb{R}^n$, $F: D \to \mathbb{R}^n$ is a continuous function, $x_0 \in D$ is a vector, A is a nonsingular $n \times n$-matrix, and

$$\delta_0 := A^{-1} F(x_0). \tag{2}$$

We suppose further that there is a constant $\kappa > 1$ such that

$$S := \{x \in \mathbb{R}^n \mid \|x - x_0\| \le \kappa \|\delta_0\|\} \subseteq D, \tag{3}$$

and a nonnegative vector $c \in \mathbb{R}^n$ such that, for some monotone norm $\|\cdot\|$

$$|F(x) - F(x_0) - A(x - x_0)| \le \|\delta_0\| c \quad \text{for all } x \in S. \tag{4}$$

Then the following theorem holds.

Theorem

If the vector

$$b := |A^{-1}| c \tag{5}$$

satisfies the condition

$$\|b\| \le \kappa - 1 \tag{6}$$

then $F(x)$ has at least one zero in S, and any such zero $\hat{x}$ satisfies

$$(2 - \kappa)\|\delta_0\| \le \|\hat{x} - x_0\| \le \kappa \|\delta_0\| . \tag{7}$$

Proof. We show that the map $\phi: D \to \mathbb{R}^n$ defined by

$$\phi(x) := x - A^{-1} F(x)$$

maps S into itself. Indeed, if $x \in S$ then

$$\begin{aligned} |\phi(x) - x_0| &= |x - x_0 - A^{-1} F(x)| \\ &= |-A^{-1}(F(x) - F(x_0) - A(x - x_0)) - A^{-1} F(x_0)| \\ &\le |A^{-1}| \cdot \|\delta_0\| c + |A^{-1} F(x_0)| \\ &= \|\delta_0\| b + |\delta_0| \end{aligned}$$

by (4), (5), and (2), hence

$$\|\phi(x) - x_0\| \le \|\delta_0\| \|b\| + \|\delta_0\| \le \kappa \|\delta_0\|$$

since $\|\cdot\|$ is monotone and by (6). Hence $\phi(x) \in S$. By Brouwer's fixed point theorem, S contains a fixed point $\hat{x}$ of ϕ, and clearly $\hat{x}$ satisfies $F(\hat{x}) = 0$.

Now for any zero $\hat{x} \in S$ of $F(x)$, the right hand inequality of (7) is obvious. To prove the left hand inequality, we insert $x = \hat{x}$, $F(\hat{x}) = 0$ into (4) and obtain

$$|F(x_0) + A(\hat{x} - x_0)| \le \|\delta_0\| c,$$

hence

$$\begin{aligned} |\hat{x} - x_0| &= |\delta_0 - A^{-1}(F(x_0) + A(\hat{x} - x_0))| \\ &\ge |\delta_0| - |A^{-1}||F(x_0) + A(\hat{x} - x_0)| \\ &\ge |\delta_0| - |A^{-1}| \cdot \|\delta_0\| c \\ &= |\delta_0| - \|\delta_0\| b, \end{aligned}$$

and therefore

$$\|\hat{x} - x_0\| \ge \|\delta_0\| - \|\delta_0\| \, \|b\| \ge (2 - \kappa)\|\delta_0\| \quad . \tag{8}$$

This completes the proof. □

To analyze the quality of the bound and the influence of the various quantities involved we adopt again the heuristic attitude of Section 1. Two important questions arise immediately:

(i) How restrictive is the condition (6) needed for the validity of (7)?

(ii) How good are the bounds (7)?

Obviously, (6) is satisfied if either c and hence b is sufficiently small, or if κ is sufficiently large.

Suppose first that κ is fixed. How small may we choose c? To answer this, we assume that $F(x)$ has a Fréchet-derivative $F'(x_0)$ in x_0, and $A = F'(x_0)$. Then $|F(x) - F(x_0) - A(x - x_0)| = o(\|x - x_0\|) = o(\|\delta_0\|)$ for $x \in S$, hence we may choose c arbitrarily small provided that $\|\delta_0\|$ is small enough. If the second derivative exists, too, then we even have $|F(x) - F(x_0) - A(x - x_0)| = O(\|\delta_0\|^2)$, hence $c = O(\|\delta_0\|)$. If A is close to $F'(x_0)$ then c has to be increased by a correction of order $\|A - F'(x_0)\|$. Thus we can satisfy (6) provided that δ_0 is sufficiently small (i.e. x_0 is sufficiently close to a zero $\hat{x}$) and A is sufficiently close to $F'(x_0)$.

Now we suppose that c is kept fixed. If $F(x)$ is differentiable then $|F(x) - F(x_0) - A(x - x_0)|$ will grow at least like a multiple of $\|x-x_0\|$, hence (4) implies that $\|x-x_0\|$ may not increase indefinitely. Therefore, the bounding factor κ in (3) must be kept sufficiently small.

Thus, if we increase κ, the vector c will increase as well. Unless F(x) is linear or almost linear, c will increase much faster than κ (for example $c = O(\kappa^2)$ if the second derivative exists). Therefore, κ has to be kept fairly small.
Of course there must be cases when no choice of c and κ will make (6) valid since F(x) may be nonzero in every set S of shape (3).

To estimate the accuracy of the bound (7) we compare the lower and upper bound. If $\kappa < 2$, then (7) shows that the upper bound overestimates the true error by a factor of at most

$$\mu = \frac{\kappa}{2-\kappa} ;$$

Thus for $\kappa = 3/2$ the true error is catched within a factor of $\mu = 3$. If $\kappa > 2$, an overestimation factor of

$$\mu' = \frac{\kappa}{1-\|b\|}$$

can be derived from (8) provided that $\|b\| < 1$. But in this case (6) is satisfied for some $\kappa < 2$ as well. Hence from the point of view of accurate estimation, there is no incentive to use a factor $\kappa > 2$. On the other hand, there are cases when (6) can be satisfied with some $\kappa > 2$ but not with $\kappa \leq 2$ (see Section 4).

Remark

The argument leading to (8) can be adapted to give the componentwise error bound

$$|\hat{x} - x_1| \leq \|\delta_0\| b \qquad \text{for } x_1 := x_0 - \delta_0 . \tag{9}$$

Thus, if the entries of b differ much in magnitude, it can be expected that some components of x_1 are much more accurate than others. (This may be an indication that the problem is badly scaled).

3. Practical aspects

In this section we discuss the choice of the parameters and the practical determination of the bound. For the sake of a simple presentation we assume that F(x) is (Fréchet-) differentiable.

<u>3.1. Choice of x_0 and A.</u> For good bounds, x_0 should be an approximation of a zero of F(x). From (6) we see that this essentially means that δ_0 is kept small. Thus it is sensible to determine x_0 from a quasi-Newton iteration.

$$z_0 \text{ suitable}, \; z_{i+1} := z_i - \sigma_i A_i^{-1} F(z_i), \; A_i \approx F'(z_i). \tag{10}$$

If $A_m^{-1} F(z_m)$ is sufficiently small we put

$$x_0 := z_m, \; A := A_m, \; \delta_0 := A_m^{-1} F(z_m). \tag{11}$$

But it should be kept in mind that the bound (7) does not depend on this choice of x_0 and A. For example, x_0 might be the solution of a previously solved near by system $\tilde{F}(x) = 0$, or (in particular if $F'(x_0)$ is an M-matrix), A might be the lower triangle of $F'(x_0)$.

<u>3.2. Choice of κ.</u> We already commented on the need of keeping κ small, preferably $\kappa < 2$. If x_0 is indeed close to a zero of F(x), and $A = F'(x_0)$ then $\kappa = 3/2$ works quite well since $c = O(\|\delta_0\|)$. But if $F'(x_0)$ is ill-conditioned, or if A is only a crude (e.g. triangular) approximation of $F'(x_0)$ then κ should be taken larger, e.g. $\kappa = 10$. If this does not work, a better approximation for $F'(x_0)$ or a better approximation x_0 of the zero is required.

<u>3.3. Determination of c.</u> There are various ways of finding a vector c satisfying (4).
Perhaps the most straightforward approach is the following. Suppose that the expression

$$F(x_0 + w) - F(x_0) - Aw \tag{12}$$

which comes from (4) by replacing x by x_0+w can be rewritten into an analytically equivalent expression each term of which is of order O(w) or smaller. The resulting expression can be bounded by eliminating w with $\|w\| \leq \kappa \|\delta_0\|$, using inequalities like the triangle inequality. If this is possible, it gives very good values for c; an example is given in Section 5.
Alternatively, the expression can be evaluated in interval arithmetic (see e.g. Moore [8]), using in place of w an interval vector containing all w with $\|w\| \leq \kappa \|\delta_0\|$. Interval arithmetic has the advantage that rounding errors are automatically taken into account (cf. 3.5), but at present most interval arithmetic implementations

are rather slow (cf., however, Moore [9]).
Another approach makes use of a technique of Alefeld [1], and can be applied if $F(x)$ is a so-called polynomial operator. Alefeld shows how to find (with interval arithmetic) two matrices $\underline{A}$ and $\overline{A}$ such that for every $x \in S$,

$$F(x) - F(x_0) = A_x(x - x_0) \tag{13}$$

for a suitable matrix A_x with

$$\underline{A} \le A_x \le \overline{A}; \tag{14}$$

then (4) amounts to finding a vector c such that

$$|(A_x - A)(x - x_0)| \le \|\delta_0\| c \quad \text{for all } x \in S. \tag{15}$$

If a scaled ℓ_∞ - norm is used,

$$\|x\| = \|x\|_u := \max_{1 \le i \le n} \frac{|x_i|}{|u_i|} = \text{Min}\,\{\alpha \ge 0 \mid |x| \le \alpha u\}, \tag{16_∞}$$

where $u = (u_1, \ldots, u_n) > 0$, then $x \in S$ implies $|x-x_0| \le \kappa \|\delta_0\| u$, whence $|(A_x - A)(x - x_0)| \le \kappa \|\delta_0\| \max(|\overline{A} - A|u, |\underline{A} - A|u)$; therefore (15) holds with

$$c := \kappa . \max(|\overline{A} - A|u, |\underline{A} - A|u). \tag{17_∞}$$

A similar, slightly more complicated argument shows that for the Euclidean norm

$$\|x\| = \|x\|_2 = \sqrt{x^T x} \tag{16_2}$$

we can satisfy (15) with

$$c := \kappa . (\nu_1, \ldots, \nu_n)^T, \tag{17_2}$$

where ν_i is the maximum of the Euclidean norms of the i-th rows of $\overline{A}$-A and $\underline{A}$-A.
Slightly less accurate estimates are obtainable from lower and upper bounds of the derivative,

$$\underline{A} \le F'(x) \le \overline{A} \quad \text{for all } x \in S; \tag{18}$$

then (13) and (14) hold with

$$A_x = \int_0^1 F'(x_0 + t(x - x_0))\, dt,$$

whence again (17) can be used to compute c.
A further possibility to get an expression for c is to use bounds for the norm of the second derivative $F''(x)$; we do not recommend the use of $F''(x)$ and hence give no details (but cf. Section 4).

<u>3.4. Determination of b.</u> We assume that a triangular decomposition

$$A = LR \tag{19}$$

of A is already available (from the computation of δ_0). The straight-forward way to compute $b=|A^{-1}|c$ is to form A^{-1}, take absolute values, and multiply with c. The formation of A^{-1} at least triples the work needed in (19), but if A is sparse the work is much more (e.g. for tridiagonal A:O(n) for (19), $O(n^2)$ for A^{-1}). Thus we look for ways to avoid the formation of A^{-1}.

The simplest case is when A is an M-matrix; then A^{-1} is nonnegative, and b can be computed as the solution of the doubly triangular system

$$LRb = c. \tag{20}$$

In general, if A^{-1} has not a constant sign, b cannot be formed without explicit knowledge of A^{-1}. Fortunately we only need an upper bound for the norm of b, hence the computation of an upper bound $\bar{b}$ of b is sufficient. It is shown in Neumaier [16] that such a bound is given by the solution of the system

$$\langle L\rangle\langle R\rangle\, \bar{b} = c, \tag{21}$$

where the <u>Ostrowski operator</u> $\langle\cdot\rangle$ replaces a matrix $M=(m_{ik})$ by

$$\langle M\rangle := (\bar{m}_{ik}) \text{ with } \bar{m}_{ik} = \begin{cases} |m_{ik}| & \text{if } i=k \\ -|m_{ik}| & \text{otherwise.} \end{cases} \tag{22}$$

If we write

$$\beta := \|b\|\ ,\ \bar{\beta} := \|\bar{b}\|$$

then $\beta \leq \bar{\beta}$, hence the condition

$$\bar{\beta} \leq \kappa - 1 \tag{23}$$

implies the hypothesis of our theorem.

If A is an M-matrix then the results of [6] imply that $\langle L\rangle=L$ and $\langle R\rangle=R$ whence $\bar{\beta}=\beta$. For other matrices, it is of critical importance to known how much $\bar{\beta}$ overestimates β. Examples show that $\bar{\beta}/\beta$ can grow exponentially with the dimension n, but for diagonally dominant matrices the situation is much better, see [10]. As an illustration, the following table gives the range of $\log_2(\bar{\beta}/\beta)$ for certain dense 20x20 matrices (band matrices behave slightly better). We chose constant entries for c, and randomly generated matrices B with unit diagonal and off-diagonal entries uniformly distributed in [-max,max]. For each value of max $\in$ {5.0, 1.0, 0.25, 0.1}, 40 such matrices were generated, and $\bar{\beta}/\beta$ was calculated for A=B, A=|B|, and A = $\langle B\rangle$, respectively.

max	5.0	1.0	0.25	0.1
A = B	15...21	12...18	2...4	$\le$ 1.5
A = |B|	16...21	12...19	2...4	$\le$ 1.5
A = <B>	15...23	9...16	7...12	$\le$ 0.5

Table 1. Range of $\log_2 (\bar{\beta}/\beta)$

An estimate for β which usually is <u>smaller</u> than β can be obtained by using $\beta = \|b\| \le \|A^{-1}\| \|c\|$ and replacing $\|A^{-1}\|$ by a <u>lower</u> bound for $\|A^{-1}\|$. Several such lower bounds have been proposed, all based on $\|A^{-1}\| \ge \|x\| / \|Ax\|$, with x suitably chosen to maximize the lower bound (see [3], [4], [11]). While such an estimate $\underline{\beta}$ cannot be used for bounding the error $\|\hat{x}-x_0\|$, it is of value in <u>assessing</u> whether a computed $\bar{\beta}$ is a realistic bound for β.

3.5. Control of rounding errors. Due to rounding errors, the quantities δ_0, c, b, and β will usually be calculated only approximately. Then the bounds (7) and (9) are valid only approximately, and if $\|b\|$ is close to $\kappa-1$, the test (6) on which (7) depends may be spoilt by rounding errors. Nevertheless, if A is not ill-conditioned, $\|b\|$ and $\|\delta_0\|$ will have the right order of magnitude, whence

$$\|b\| \ll \kappa - 1 \tag{6a}$$

is still an indication that (7) holds.

If reliable bounds are required we may use rounded interval arithmetic (Moore [8]) which automatically takes account of the rounding errors made by working with the intervals of uncertainly of each variable. The solution of linear interval equations is easiest if A is diagonally dominant or at least an H- matrix, for then the interval version of Gauss elimination can be carried out without problems (cf. Alefeld [1]); moreover the system (21) can be programmed in the equivalent but simpler form

$$[-\bar{b},\bar{b}] := \text{solution of } LRz = [-c,c]. \tag{21a}$$

4. The Kantorovic bound

In this section we compare our bounds with the a priori bound which is part of the famous theorem of Kantorovic. This theorem gives conditions which guarantee that Newton's iteration converges to a zero of a function F(x). As a byproduct, the following bounds for a zero are obtained (cf. e.g. Ortega and Rheinboldt [12]).

Proposition (Kantorovic)
Let D be a subset of $\mathbb{R}^n$, and suppose that the function $F:D \to \mathbb{R}^n$ has in D a continuous Fréchet derivative $F'(x)$. Assume that for some $x_0 \in D$, $F'(x_0)$ is nonsingular, and

$$||F'(x_0)^{-1} F(x_0)|| \leq \alpha \neq 0, \tag{24}$$

$$||F'(x_0)^{-1}|| \leq \beta ,$$

$$||F'(x) - F'(y)|| \leq \gamma \, ||x - y|| \quad \text{for } x, y \in S_0,$$

where S_0 is a suitable convex subset of D. If

$$h := 2\alpha\beta\gamma \leq 1, \tag{25}$$

and

$$S := \{x \in \mathbb{R}^n \mid ||x - x_0|| \leq r := \frac{2\alpha}{1+\sqrt{1-h}}\} \subseteq S_0 \tag{26}$$

then F(x) has a unique zero in S. □

Remarks. 1. There is also an affine invariantive form of the proposition, see e.g. Miel [7]. But in practical applications this simply amounts to applying the preceding to BF(x) in place of F(x), with a suitable preconditioning matrix B.

2. Usually, the constant γ is determined as a bound of the norm of the second derivative $F''(x)$ in S_0.

3. γ usually depends on the choice of S_0. Since in (26), $\alpha < r \leq 2\alpha$; it is sensible to choose

$$S_0 := \{x \in \mathbb{R}^n \mid ||x - x_0|| \leq \kappa\alpha\} . \tag{27}$$

with a parameter κ satisfying

$$1 < \kappa \leq 2; \tag{28}$$

of course, we must assume that $S_0 \subseteq D$ (note that Rall [14], Theorem 24.1. recommends to use (27) with $\kappa = 2$). With this choice of S_0,

(25) and (26) are equivalent with

$$h := 2\alpha\beta\gamma \leq 4(\kappa - 1)/\kappa^2 . \tag{29}$$

4. The relation

$$\|F(x)-F(x_0)-F'(x_0)(x-x_0)\| \leq \frac{1}{2}\gamma\,\|x-x_0\|^2 \quad \text{for } x,y \in S_0 \tag{30}$$

is a consequence of (24); see [12].

Now assume that $\|\ \| = \|\cdot\|_u$ is a scaled ℓ_∞-norm. Then, with S_0 as in (27), we find from (30) that

$$|F(x) - F(x_0) - F'(x_0)(x - x_0)|\, u \leq \frac{1}{2}\gamma\kappa^2\alpha^2\, u. \tag{31}$$

Hence we are in the situation of Section 2 with

$$A = F'(x_0),\ \|\delta_0\| = \alpha,\quad c = \frac{1}{2}\alpha\gamma\kappa^2\, u,$$

and we find $\|b\|=\| |A^{-1}|c\|_u = \frac{1}{2}\alpha\gamma\kappa^2\| |A^{-1}|u\|_u = \frac{1}{2}\alpha\gamma\kappa^2\|A^{-1}\|_u \leq \frac{1}{2}\alpha\gamma\kappa^2\beta = \frac{1}{4}h\kappa^2$.

We see that (6) and (29) are equivalent inequalities. Hence, for scaled ℓ_∞-norms, the a priori Kantorovic bound is equivalent to our theorem, when c is computed from the relation (30).

Now we show by an example (in dimension n=1) that our theorem may produce a bound when Kantorovic's bound does not work. Consider

$$F(x) = x^3 + 12x + 12, \qquad D = \mathbb{R}$$

$$x_0 = 0,\ A = F'(x_0) = 12.$$

Then Kantorovic's constants are

$$\alpha = 1,\ \beta = \frac{1}{12},\ \gamma = \sup_{x,y \in S_0} 3|x + y| \geq 6 \sup_{x \in S_0}|x|.$$

Condition (25) implies that $\gamma \leq 6$, hence $\sup_{x \in S_0} |x| \leq 1 = \alpha$, whence

$$S_0 \subseteq \{x \in \mathbb{R}\mid \|x - x_0\| \leq \alpha\},$$

and (26) cannot be satisfied. Therefore, no bound is obtained. On the other hand, our theorem, applied with $\kappa = 3/2$ gives

$$\delta_0 = A^{-1}F(x_0) = 1,$$

$$S = \{x \in \mathbb{R} \mid\ |x| \leq 3/2\},$$

$$|F(x) - F(x_0) - A(x - x_0)| = |x^3| \leq 27/8,$$

whence

$$c = 27/8,\ b = |A^{-1}|c = 9/32 \leq 1/2 = \kappa - 1.$$

Therefore, $F(x)$ has a zero $\hat{x}$ with $|\hat{x}| \leq {}^3/2$. Note that (9) gives the sharper estimate $|\hat{x}+1| \leq {}^9/32$, whence $-1.3 < \hat{x} < -0.7$. In fact, we have $\hat{x} \approx -0.932441$.

From the relationship to Kantorovic's bound it again appears that it is sensible to choose $\kappa \leq 2$. But there are cases when a value of $\kappa > 2$ is useful. For example, choose $t > 0$ and put

$$F(x) = \begin{cases} e^x - 1 & \text{if } x \leq t, \\ e^t - 1 + e^t(x - t) & \text{if } x > t. \end{cases} \tag{32}$$

Then $F(x)$ has a continuous derivative

$$F'(x) = e^{\min(x,t)}.$$

The unique zero of $F(x)$ is $\hat{x} = 0$. Hence if we take

$$x_0 := t, \quad A := F'(x_0) = e^t,$$

so that

$$\delta_0 = 1 - e^{-t} \quad (> 0),$$

the set (3) contains the zero only if $\kappa \, \|\delta_0\| > t$, or

$$\kappa \geq t/(1 - e^t). \tag{33}$$

Now suppose we choose κ according to (33), so that

$$r := \kappa \, \delta_0 \geq t.$$

Since

$$F(x_0 + w) - F(x_0) - Aw = \begin{cases} e^t(e^w - w - 1) & \text{if } w < 0, \\ 0 & \text{if } w \geq 0, \end{cases}$$

is nonnegative and decreasing, it takes its maximal absolute value for the smallest admissible w. Therefore, (4),(5) hold with

$$c = e^t(e^{-r} + r - 1)/\delta_0,$$

$$b = (e^{-r} + r - 1)/\delta_0 = \kappa - (1 - e^{-r})/(1 - e^{-t}) \leq \kappa - 1.$$

By our theorem, the bound

$$|\hat{x} - x_0| \leq \kappa \, \delta_0 \tag{34}$$

is valid for every κ satisfying (33).

<u>Remarks</u>. 1. If $t \geq 2$ then (33) implies $\kappa > 2$, whence no value of $\kappa \leq 2$ can be used to bound the zero.

2. If x_0 is "close" to the zero $\hat{x}$ then t is small, and we may take κ small as well; note that $t/(1-e^{-t}) = 1+\frac{1}{2}t+O(t^2)$.

3. The example is unusual in that <u>every</u> large κ gives a bound. This is due to the fact that the function $F(x)$ is asymptotically linear for large x.

4. If κ is chosen optimally (equality in (33)) then the bound (34) gives the error exactly (equality in the bound). Although this does not generalize to arbitrary functions $F(x)$, it indicates the quality of the estimate.

5. Numerical example

As an illustration, we consider the nonlinear bounding value problem ("minimal surface equation")

$$(1 + u_y^2)\, u_{xx} - 2u_x u_y u_{xy} + (1 + u_x^2)\, u_{yy} = 0 \quad \text{in } D, \qquad (35)$$

$$u(x,y) = g(x,y) \text{ on the boundary } \partial D.$$

We look at the problem when D is the square,

$$D = \{(x,y) \mid 0 \le x \le 1,\ 0 \le y \le 1\},$$

replace the square by a grid of $(N+1)^2$ points,

$$\left(\frac{i}{N}, \frac{k}{N}\right), \quad i,k \in \{0,1,\dots,N\},$$

and approximate the function u and its derivative in these points as follows:

$$u\left(\frac{i}{N}, \frac{k}{N}\right) \approx v_{ik},$$

$$u_x\left(\frac{i}{N}, \frac{k}{N}\right) \approx \frac{N}{2}\, \Delta_x v_{ik}, \quad u_y\left(\frac{i}{N}, \frac{k}{N}\right) \approx \frac{N}{2}\, \Delta_y v_{ik},$$

$$u_{xx}\left(\frac{i}{N}, \frac{k}{N}\right) \approx N^2\, \Delta_{xx} v_{ik}, \quad u_{yy}\left(\frac{i}{N}, \frac{k}{N}\right) \approx N^2\, \Delta_{yy} v_{ik},$$

$$u_{xy}\left(\frac{i}{N}, \frac{k}{N}\right) \approx \frac{N^2}{4}\, \Delta_{xy} v_{ik},$$

where

$$\Delta_x v_{ik} = v_{i+1,k} - v_{i-1,k}, \qquad \Delta_y v_{ik} = v_{i,k+1} - v_{i,k-1},$$

$$\Delta_{xx} v_{ik} = v_{i+1,k} - 2v_{ik} + v_{i-1,k}, \quad \Delta_{yy} v_{ik} = v_{i,k+1} - 2v_{i,k} + v_{i,k-1},$$

$$\Delta_{xy} v_{ik} = v_{i+1,k+1} - v_{i+1,k-1} - v_{i-1,k+1} + v_{i-1,k-1}.$$

The boundary conditions give

$$v_{ik} = g\left(\frac{i}{N}, \frac{k}{N}\right) \text{ if } \ i \in \{0,N\} \ \text{ or } k \in \{0,N\},$$

and the remaining v_{ik} are arranged as an $(N-1)^2$-dimensional block vector

$$v = (v_1, \dots, v_{N-1})^T; \quad \text{with } v_i = (v_{i1}, \dots, v_{i,N-1})^T.$$

At a fixed interior point $\left(\frac{i}{N}, \frac{k}{N}\right)$, the differential equation is replaced by the discrete equation

$$0 = F_{ik}(v) = (\alpha+\delta_y^2)\delta_{xx} - \frac{1}{2}\delta_x\delta_y\delta_{xy} + (\alpha+\delta_x^2)\delta_{yy}, \qquad (36_{ik})$$

where

$$\alpha = 4/N^2,$$

$$\delta_x = \Delta_x v_{ik},\ \delta_y = \Delta_y v_{ik},\ \text{etc.}$$

For a small vector w with

$$\|w\|_\infty \le \omega, \qquad (37)$$

$$w_{ik} = 0 \quad \text{if } i \in \{0,N\} \quad \text{or } k \in \{0,N\}, \qquad (38)$$

the quantities

$$\varepsilon_x = \Delta_x v_{ik},\ \varepsilon_y = \Delta_y v_{ik},\ \text{etc.}$$

are bounded by

$$|\varepsilon_x| \le 2\omega,\ |\varepsilon_y| \le 2\omega,$$

$$|\varepsilon_{xx}| \le 4\omega,\ |\varepsilon_{xy}| \le 4\omega,\ |\varepsilon_{yy}| \le 4\omega.$$

Now

$$F_{ik}(v+w) = (\alpha+(\delta_y+\varepsilon_y)^2)(\delta_{xx}+\varepsilon_{xx}) - \frac{1}{2}(\delta_x+\varepsilon_x)(\delta_y+\varepsilon_y)(\delta_{xy}+\varepsilon_{xy})$$
$$+(\alpha+(\delta_x+\varepsilon_x)^2)(\delta_{yy}+\varepsilon_{yy}).$$

We extract from this the part linear in the epsilons and obtain a formula for the derivative:

$$[F'(v)w]_{ik} = (\alpha+\delta_y^2)\varepsilon_{xx} - \frac{1}{2}\delta_x\delta_y\varepsilon_{xy} + (\alpha+\delta_x^2)\varepsilon_{yy}$$
$$+(2\delta_x\delta_{yy} - \frac{1}{2}\delta_y\delta_{xy})\varepsilon_x + (2\delta_y\delta_{xx} - \frac{1}{2}\delta_x\delta_{xy})\varepsilon_y. \qquad (39_{ik})$$

If we take differences we find

$$F_{ik}(v+w) - F_{ik}(v) - [F'(v)w]_{ik} =$$
$$= \varepsilon_y^2(\delta_{xx}+\varepsilon_{xx}) - \frac{1}{2}\varepsilon_x\varepsilon_y(\delta_{xy}+\varepsilon_{xy}) + \varepsilon_x^2(\delta_{yy}+\varepsilon_{yy})$$
$$+2\delta_y\varepsilon_y\varepsilon_{xx} - \frac{1}{2}(\delta_x\varepsilon_y+\varepsilon_x\delta_y)\varepsilon_{xy} + 2\delta_x\varepsilon_x\varepsilon_{yy},$$

hence

$$|F(v+w) - F(v) - F'(v)w|_{ik} \le 4\omega^2(e_{ik} + 10\omega), \qquad (40_{ik})$$

where

$$e_{ik} = |\delta_{xx}| + \frac{1}{2}|\delta_{xy}| + |\delta_{yy}| + 5|\delta_x| + 5|\delta_y|. \qquad (41_{ik})$$

Now, if v is an approximation to a zero of the system $F(v)=0$ given by (36_{ik}) for $i,k \in \{1,\ldots,N-1\}$ then we may apply our theorem with $A=F'(v)$; an explicit (block tridiagonal) matrix expression is availab-le from (39_{ik}) by expansion to linear combinations of the $v_{i+j,k+\ell}$, $j,\ell \in \{0,1,-1\}$. Once $\delta_0=A^{-1}F(v)$ and

$$\omega = \kappa \|\delta_0\|$$

are computed, c is available from (40) as

$$c_{ik} = 4\kappa\omega(e_{ik} + 10\omega); \tag{42_{ik}}$$

if the e_{ik} are computed simultaneously with $F(v)$ and $F'(v)$, little extra effort is necessary.

As a numerical example we consider the problem (35) on the unit square D, with

$$g(x,y) = \sqrt{xy(2-xy)}.$$

We take $g(x,y)$ as initial approximation for $u(x,y)$. For $N=8$, we obtain the following initial block vector $v = (v_{ik})$:

.0000	.0000	.0000	.0000	.0000	.0000	.0000	.0000	.0000
.0000	.1761	.2480	.3026	.3480	.3875	.4227	.4547	.4841
.0000	.2480	.3480	.4227	.4841	.5367	.5830	.6242	.6614
.0000	.3026	.4227	.5113	.5830	.6433	.6953	.7407	.7806
.0000	.3480	.4841	.5830	.6614	.7262	.7806	.8268	.8660
.0000	.3875	.5367	.6433	.7262	.7929	.8472	.8914	.9270
.0000	.4227	.5830	.6953	.7806	.8472	.8992	.9391	.9682
.0000	.4547	.6242	.7407	.8268	.8914	.9391	.9721	.9922
.0000	.4841	.6614	.7806	.8660	.9270	.9682	.9922	1.0000

We use $\kappa = 1.5$ and find after i Newton steps

$\overline{\beta} = 630.5$, $\omega = 0.3270$, for $i = 0$,

$\overline{\beta} = 0.4788$, $\omega = 0.00065$, for $i = 6$,

$\overline{\beta} = 0.1316$, $\omega = 0.00018$, for $i = 7$.

Thus the error estimate is unreliable for $i = 0$ ($\overline{\beta} \gg \kappa-1 = 0.5$), probably reliable for $i = 6$ ($\overline{\beta} \approx \kappa-1$), and reliable for $i = 7$ ($\overline{\beta} \ll \kappa-1$). Hence the discretized solution agrees with the approximation for $i = 7$.

.0000	.0000	.0000	.0000	.0000	.0000	.0000	.0000	.0000
.0000	.0219	.0446	.0690	.0972	.1333	.1860	.2780	.4841
.0000	.0445	.0903	.1391	.1941	.2615	.3516	.4812	.6614
.0000	.0690	.1391	.2121	.2914	.3821	.4908	.6235	.7806
.0000	.0972	.1942	.2914	.3910	.4958	.6088	.7317	.8660
.0000	.1333	.2615	.3821	.4958	.6047	.7110	.8169	.9270
.0000	.1861	.3517	.4909	.6088	.7110	.8017	.8852	.9682
.0000	.2782	.4812	.6235	.7317	.8169	.8852	.9413	.9922
.0000	.4841	.6614	.7806	.8660	.9270	.9682	.9922	1.0000

within two units of the last decimal. Of course, the discretization error is not included in our bound.

Our second example is the catenoid

$$u(x,y) = \cosh^{-1} \sqrt{(x+1)^2+(y-0.5)^2}, \tag{44}$$

which is a solution of the minimal surface equation. We take it as initial approximation for the discretized equation with $N = 8$ on the unit square. The computed error bound is now a bound for the discretization error of this particular problem. We obtain (for $\kappa = 1.5$)

$$\overline{\beta} = 3.2265, \qquad \omega = 0.0062,$$

hence no reliable bound. But after two Newton steps, we have

$$\overline{\beta} = 0.0987, \qquad \omega = 0.0002;$$

inspection shows that the two approximations differ by at most 0.0045, whence the first estimate is also valid.

Finally, we check an example on the square $[0,20] \times [0,20]$ given by Douglas [5], who used essentially the same discretization and calculated the following approximation (in 1928, by hand) using coordinate relaxation:

.0	9.7	13.2	15.6	17.3	18.5	19.4	19.8	20.0
.0	5.4	9.4	12.4	14.6	16.3	17.8	18.9	19.8
.0	3.5	6.8	9.6	12.0	14.1	16.0	17.8	19.4
.0	2.5	4.9	7.4	9.7	11.9	14.1	16.3	18.5
.0	1.8	3.7	5.5	7.5	9.7	12.0	14.6	17.3
.0	1.3	2.6	4.0	5.5	7.4	9.6	12.4	15.6
.0	.8	1.7	2.6	3.7	4.9	6.8	9.4	13.2
.0	.4	.8	1.3	1.8	2.5	3.5	5.4	9.7
.0	.0	.0	.0	.0	.0	.0	.0	.0

For $\kappa = 1.5$ we obtain

$$\overline{\beta} = 38.3463, \qquad \omega = 0.8544.$$

The correctly rounded solution

.0	9.7	13.2	15.6	17.3	18.5	19.4	19.8	20.0
.0	5.6	9.6	12.5	14.6	16.3	17.7	18.8	19.8
.0	3.7	7.0	9.8	12.2	14.2	16.0	17.7	19.4
.0	2.7	5.2	7.6	9.9	12.1	14.2	16.3	18.5
.0	1.9	3.9	5.8	7.8	9.9	12.2	14.6	17.3
.0	1.4	2.8	4.2	5.8	7.6	9.8	12.5	15.6
.0	.9	1.8	2.8	3.9	5.2	7.0	9.6	13.2
.0	.4	.9	1.4	1.9	2.7	3.7	5.6	9.7
.0	.0	.0	.0	.0	.0	.0	.0	.0

has error parameters

$$\overline{\beta} = 2.6193, \qquad \omega = 0.0742.$$

In both cases we have $\overline{\beta} > \kappa - 1$ although the true maximal error is $< \omega$. This is partially explained by the fact that the Newton iteration converges very slowly: For the first 25 iterations starting from Douglas' approximation, the convergence is only linear, with factor 0.7. The first iterate with $\overline{\beta} < \kappa - 1$ is the 16th, with

$$\overline{\beta} = 0.2808, \qquad \omega = 0.0082$$

(this allows checking the validity of the other two bounds). Thus we have here a very nonlinear problem.

All examples were calculated on a UNIVAC 1110. The time for one Newton iteration was 0.13 sec (without error information) resp. 0.16 sec (with error information)

Use was made of (21) to compute $\overline{\beta}$, although the matrices A were neither M-matrices nor diagonally dominant. Perhaps this accounts for the fact that in the examples ω was a bound even when $\overline{\beta}$ was too large.

Acknowledgment I want to thank Prof. Collatz for his critical remarks, and Prof. Törnig for suggesting the minimal surface equation as an interesting application.

References

1. G. Alefeld, Über die Durchführbarkeit des Gaußschen Algorithmus bei Gleichungen mit Intervallen als Koeffizienten, Computing , Suppl. 1 (1977), 15-19.

2. G. Alefeld, Bounding the slope of polynomial operators and some applications, Computing 26 (1981), 227-237.

3. A.L. Cline, C.B. Moler, G.W. Stewart, and J.H. Wilkinson, An estimate for the condition number of a matrix, SIAM J.Numer.Anal. 16 (1979), 368-375.

4. J.J. Dongarra, J.R. Bunch, C.B. Moler, and G.W. Stewart, LINPACK Users Guide. SIAM, Philadelphia 1979.

5. J. Douglas, A method of numerical solution of the problem of Plateau, Annals of Math. (2) 29 (1928), 180-188.

6. M. Fiedler and V. Ptak, On matrices with nonpositive off-diagonal elements and positive principal minors, Czech. Math.J. 12 (1962), 382-400.

7. G. Miel, An updated version of the Kantorovich theorem for Newton's method, Computing, to appear.

8. R.E. Moore, Methods and Applications of Interval Analysis, SIAM Publications, Philadelphia 1979.

9. R.E. Moore, New results on nonlinear systems. In: Interval Mathematics 1980 (Ed. K. Nickel), Acad. Press, New York - London 1980, pp. 165-180.

10. A. Neumaier, Hybrid norms, the Ostrowski operator, and bounds for solutions of linear equations, to appear.

11. D.P. O'Leary, Estimating matrix condition numbers, SIAM J. Sci.Stat.Comput. 1 (1980), 205-209.

12. J.M. Ortega and W.C. Rheinboldt, Iterative solution of nonlinear equations in several variables. Acad. Press, New York - London 1970.

13. A.M. Ostrowski, über die Determinanten mit überwiegender Hauptdiagonale, Comm.Math.Helv. 10 (1937), 69-96.

14. L.B. Rall, Computational solution of nonlinear operator equations. Wiley, New York - London 1969.

DAS AUFLÖSUNGSVERHALTEN VON NICHTLINEAREN FIXMENGEN-SYSTEMEN

Karl Nickel

Institut für
Angewandte Mathematik
der Universität

D 7800 Freiburg i.Br.

GERMANY

Zusammenfassung: Es sei F eine Abbildung eines passend gewählten n-dimensionalen metrischen Mengenraumes in sich. Die Funktion F möge dort einerseits einer Lipschitz-Bedingung und andererseits einer Durchmesser-Bedingung genügen. Es werden Systeme der Gestalt

$$X = F(X) + R \qquad (1)$$

betrachtet. Es werden 6 äquivalente Bedingungen angegeben, die sämtlich notwendig und hinreichend sind für die eindeutige Auflösbarkeit von (1). Eine dieser Bedingungen ist die Konvergenz des Iterations-Verfahrens.

Abstract: Let F be a mapping of a certain n-dimensional domain of sets. Assume that F satisfies there a Lipschitz condition together with a diameter condition. Systems

$$X = F(X) + R \qquad (1)$$

are studied. In what follows 6 equivalent conditions are presented which are necessary and sufficient for the unique solvability of (1). One of these conditions is the convergence of the iteration method applied to (1).

1. Einleitung.

Es sei G ein Mengenbereich, versehen mit der üblichen Ordnungsrelation $\subseteq$ und mit einer Addition $+: G \times G \to G$. Auf G existiere eine n-dimensionale Pseudometrik $|\cdot,\cdot|: G \times G \to \mathbb{R}^n_+$ und G sei vollständig bezüglich dieser Metrik.

In der folgenden Arbeit sollen nichtlineare Fixmengengleichungen der Gestalt

$$X = F(X) + R \tag{1}$$

betrachtet werden. Dabei sei $R \in G$ gegeben und $X \in G$ gesucht. Die gegebene Abbildung $F: G \to G$ genüge auf G einer Lipschitzbedingung

$$|F(X),F(Y)| \leqq \Lambda |X,Y| \quad \text{für} \quad X,Y \in G \tag{2}$$

mit einer Lipschitzkonstanten ("Lipschitzmatrix") $\Lambda \in \mathbb{R}^{n,n}_+$.

Die folgenden Eigenschaften sind fast selbstverständlich und wohlbekannt: gilt für den Spektralradius $\rho(\Lambda)$ der Lipschitzmatrix Λ aus (2) die Ungleichung

$$\rho(\Lambda) < 1, \tag{3}$$

dann besitzt das Problem (1) eine eindeutig bestimmte Lösung $\hat{X} = \hat{X}(R) \in G$ für alle Mengen $R \in G$. Diese Lösung $\hat{X}$ läßt sich mit dem - stets konvergenten - Iterationsverfahren gewinnen. Ist zusätzlich zu (3) die Abbildung F noch inklusionsisoton und ist die Inklusion $\subseteq$ stetig bezüglich der Pseudometrik, dann hängt die Lösung $\hat{X}(R)$ von (1) isoton ab sowohl von der Menge R, als auch von der Abbildung F. Diese Eigenschaften sind in den ersten 3 Sätzen der folgenden Arbeit formuliert.

Mit (3) ist eine einfache hinreichende Existenz- und Eindeutigkeitsbedingung zu (1) gefunden. Es ist das wesentliche Ziel der nachfolgenden Untersuchungen, Bedingungen für die eindeutige Auflösbarkeit von (1) anzugeben, die sowohl notwendig als auch hinreichend sind. Man kann nicht erwarten, daß es ohne zusätzliche Voraussetzungen an die Funktion F einfache derartige Kriterien geben wird. Das zeigt schon der Sonderfall *linearer* Abbildungen F, man vergleiche dazu etwa Davis-Najfeld-Vitale [4]. Auch bei dem entsprechenden Fix*punkt*problem, wenn der Grundmengenbereich G durch einen n-dimensionalen *linearen* Raum ersetzt wird, sind die genauen Lösungsverhältnisse des Systems (1) nicht einfach zu beschreiben. Um so überraschender ist es, daß die folgende Vorgehensweise zum Ziel führt: Auf G wird ein n-dimensionaler "Durchmesser" diam: $G \to \mathbb{R}_+^n$ eingeführt. Zusätzlich zur Lipschitzbedingung (2) möge die Funktion F noch einer Durchmesserbedingung

$$\text{diam } F(X) \geqq \Lambda \text{ diam } X \quad \text{für} \quad X \in G \tag{4}$$

genügen, wobei Λ die Lipschitzmatrix aus (2) sein soll. Man betrachtet die eindeutige Lösbarkeit von (1) nicht nur für eine spezielle feste Menge R, sondern für alle Mengen $R \in G$. Dann lassen sich insgesamt 6 Aussagen angeben, die alle zu dieser eingeschränkten Lösbarkeitsaussage äquivalent sind. Erstaunlicherweise gehört auch die Bedingung (3) mit dazu! Dieses Ergebnis ist der Inhalt von Satz 4 der vorliegenden Arbeit. Dabei ist es wichtig zu betonen, daß diese Äquivalenzen nur richtig sind für Mengensysteme. Sie gelten also nicht für Punktsysteme (1), obwohl dann die Forderung (4) immer erfüllt ist wegen diam F(X) = diam X = 0.

Man kann sich nun fragen, wie die Klasse aller Abbildungen F beschaffen ist, für die die Ungleichungen (2) und (4) gelten. Man zeigt leicht, daß für lineare Intervall- oder Kreisscheiben-Gleichungssysteme die Bedingungen (2) und (4) automatisch erfüllt sind und daß damit unter anderen Satz 4 gilt. Dies konnte schon in [5] bewiesen werden. Im folgenden werden zwei einfache Beispiele für nichtlineare Abbildungen angegeben und damit gezeigt, daß sich zwar (2) und (4) teilweise entgegenstehen, aber nicht vollständig widersprechen und daß es damit weite Klassen von Problemen gibt, auf die sich die hier entwickelte Theorie anwenden läßt.

Anregungen zu dieser Untersuchung verdankt der Autor den Herren Prof. Dr. R. Krawczyk/Clausthal-Zellerfeld und Doz. Dr. A. Neumaier/ Freiburg i. Br..

2. Definitionen und Bezeichnungen.

Es sei $\mathbb{R}$ wie üblich der Körper der reellen Zahlen. Zu der natürlichen Zahl $n \geq 1$ werden mit $\mathbb{R}^n$ und $\mathbb{R}^{n,n}$ die Mengen der reellen n-Vektoren und n×n-Matrizen bezeichnet. Ist $\Lambda \in \mathbb{R}^{n,n}$, dann soll $\rho(\Lambda)$ den Spektralradius der Matrix Λ bedeuten. Mit der komponentenweisen Ordnungsrelation $\leq$ seien $\mathbb{R}^n_+$ und $\mathbb{R}^{n,n}_+$ die jeweiligen Ordnungskegel. Ungleichungen $\leq$ und $<$ sind auf $\mathbb{R}^n$ stets komponentenweise aufzufassen: gilt etwa $\alpha = (\alpha_1,\alpha_2,\ldots,\alpha_n)$, $\beta = (\beta_1,\beta_2,\ldots,\beta_n) \in \mathbb{R}^n$, so soll $\alpha < \beta$ damit $\alpha_i < \beta_i$ für $i = 1(1)n$ bedeuten. Die eindimensionale Null $0 \in \mathbb{R}$ und der Nullvektor $0 \in \mathbb{R}^n$ sollen nicht verschieden bezeichnet werden, da keine Verwechslungen zu befürchten sind. Wie üblich bedeute $I \in \mathbb{R}^{n,n}$ die Einheitsmatrix. Später wird noch an je einer Stelle eine (beliebige) Norm $\|\cdot\|: \mathbb{R}^n \rightarrow \mathbb{R}_+$ und der komponentenweise Vektorbetrag $|\cdot|: \mathbb{R}^n \rightarrow \mathbb{R}^n_+$ verwendet, der für $\alpha = (\alpha_1,\alpha_2,\ldots,\alpha_n) \in \mathbb{R}^n$ durch $|\alpha| := (|\alpha_1|,|\alpha_2|,\ldots,|\alpha_n|)$ erklärt ist.

Es sei $S = \{a,b,\ldots\}$ eine (additiv geschriebene) abelsche Gruppe mit dem Nullelement ϑ. Wie üblich bedeute $-a$ die Inverse zum Element $a \in S$; ferner werde $a-b := a+(-b)$ gesetzt. Mit $\mathbb{P}(S)$ sei die Potenzmenge über S bezeichnet; die Relationen $=$ und $\subseteq$ seien auf $\mathbb{P}(S)$ wie üblich festgelegt. Im folgenden soll das Symbol $\{a\}$ für $a \in S$ stets eine einpunktige Menge $\in \mathbb{P}(S)$ bezeichnen. Die Addition + wird von S auf $\mathbb{P}(S)$ erweitert durch die übliche Definition

$$A + B := \{a+b \mid a \in A,\ b \in B\} \tag{5}$$

für $A,B \in \mathbb{P}(S)$. Diese Operation + ist kommutativ und assoziativ, sie wird auch gelegentlich "Minkowski-Summe" genannt, siehe [4]. Mit $\Theta := \{\vartheta\}$ werde das Nullelement von $\mathbb{P}(S)$ bezeichnet. Man beachte, daß $\mathbb{P}(S)$ mit der Operation nach (5) <u>keine</u> abelsche Gruppe ist. Man setzt weiter

$$A - B := \{a-b \mid a \in A,\ b \in B\},$$

damit ist insbesondere

$$A - A = \{a-b \mid a,b \in A\}.$$

Es gilt

$$A - A = \Theta \Leftrightarrow A = \{a\} \text{ mit } a \in S.$$

Eine Abbildung $|\cdot,\cdot|: \mathbb{P}(S)\times\mathbb{P}(S) \to \mathbb{R}^n_+$ heißt (symmetrische) Pseudometrik auf $\mathbb{P}(S)$, wenn für alle $A,B,C \in \mathbb{P}(S)$ gilt

$$|A,B| = 0 \Leftrightarrow A = B,$$

$$|A,B| = |B,A|,$$

$$|A,C| \leqq |A,B| + |B,C|.$$

Man vergleiche dazu Kurepa [3] oder Collatz [1]; gelegentlich wird in der Literatur auch der Name "Vektormetrik" verwendet. Sie heißt inklusionsisoton, wenn für alle $A,B,C \in \mathbb{P}(S)$ gilt

$$A \subseteq B \subseteq C \Rightarrow |A,B| \leqq |A,C|.$$

Gilt

$$|A+C,\ B+C| = |A,B| \quad \text{für} \quad A,B,C \in \mathbb{P}(S),$$

so heißt die Pseudometrik translations-invariant.
Im folgenden wird die schwächere Bedingung

$$|A+C,\ B+C| \leqq |A,B| \tag{6}$$

vorausgesetzt, sie soll entsprechend translations-sub-invariant genannt werden.

Für die nachfolgenden Untersuchungen werde als Grundmenge eine Teilmenge $G \subseteq \mathbb{P}(S)$ zugrundegelegt. Es gelte $\Theta \in G$ und $G \setminus \Theta \neq \emptyset$. Die Menge $G = \{A,B,\ldots\}$ sei abgeschlossen bezüglich der arithmetischen Operation + gemäß (5). Ferner sei G in dem folgenden schwachen Sinne ein n-dimensionaler Raum: Auf G gebe es eine translations-sub-invariante und inklusionsisotone n-dimensionale Pseudometrik und G sei damit metrisch vollständig.

Man erklärt weiter als "Durchmesser" eines Elements $A \in G$ die Abbildung diam: $G \to \mathbb{R}^n_+$ durch

$$\operatorname{diam} A := |A-A,\ \Theta|.$$

Dieser Durchmesser genügt damit den folgenden Regeln für alle $a \in S$ mit $\{a\} \in G$ und für $A,B,C \in G$:

$$\operatorname{diam} A = 0 \Leftrightarrow A = \{a\},$$

$$\operatorname{diam}\,(A+\{a\}) = \operatorname{diam} A, \tag{7}$$

$$\operatorname{diam} A \leq \operatorname{diam}(A+B) \leq \operatorname{diam} A + \operatorname{diam} B, \tag{8}$$

$$B \subseteq C \Rightarrow \operatorname{diam}(A+B) \leq \operatorname{diam}(A+C)$$

und damit insbesondere

$$A \subseteq B \Rightarrow \operatorname{diam} A \leq \operatorname{diam} B. \tag{9}$$

Zusätzlich zu der Mengeninklusion $\subseteq$ werde auf G noch eine "starke" Inklusion $\dot{\subset}$ eingeführt durch die Setzung

$$A \dot{\subset} B :\Leftrightarrow A \subseteq B \wedge \operatorname{diam} A < \operatorname{diam} B \tag{10}$$

für alle $A,B \in G$.

3. Sätze.

Es sei $F: G \to G$ eine Abbildung, die auf G einer Lipschitzbedingung

$$|F(X),F(Y)| \leq \Lambda |X,Y| \quad \text{für} \quad X,Y \in G \tag{2}$$

mit einer Lipschitzkonstanten $\Lambda \in \mathbb{R}_+^{n,n}$ genüge. Der folgende erste Satz ist wohlbekannt:

<u>Satz 1:</u> *Die Funktion $F: G \to G$ genüge einer Lipschitzbedingung (2) und für den Spektralradius $\rho(\Lambda)$ der Lipschitzkonstanten Λ in (2) gelte*

$$\rho(\Lambda) < 1. \tag{3}$$

i) Dann hat das Fixmengenproblem

$$X = F(X) + R \tag{1}$$

für jedes Element $R \in G$ in G eine eindeutig bestimmte Lösung $\hat{X} = \hat{X}(R) \in G$.

ii) Zu $R \in G$ sei die Folge $\{Y_\nu\}$ mit $Y_\nu \in G$ durch das Iterationsverfahren

$$\left.\begin{aligned} &Y_0 \in G \\ &Y_{\nu+1} := F(Y_\nu) + R \quad \text{für} \quad \nu = 0,1,\ldots \end{aligned}\right\} \tag{11}$$

definiert. Dann gilt für alle Anfangswerte $Y_o \in G$ *stets (im Sinne der Pseudometrik)*

$$\lim_{\nu\to\infty} Y_\nu = \hat{X}.$$

Dabei ist $\hat{X} = \hat{X}(R) \in G$ *die (nach i) eindeutig bestimmte) Lösung von (1).*

<u>*Zusatz:*</u>

iii) Für das Iterationsverfahren (11) gelten die üblichen a posteriori- und a priori-Schranken

$$|\hat{X},Y_\nu| \leqq (I-\Lambda)^{-1}\Lambda|Y_\nu,Y_{\nu-1}| \leqq (I-\Lambda)^{-1}\Lambda^\nu|Y_1,Y_o|.$$

iv) Ist $\check{X}$ *eine (beliebig gefundene) Näherung zu* $\hat{X}$*, dann gilt die Fehlerschranke*

$$|\hat{X},\check{X}| \leqq (I-\Lambda)^{-1}|\check{X},F(\check{X})+R|.$$

<u>Beweis</u>: Es genügt, allein den Teil ii) zu betrachten. Wegen der Translations-sub-Invarianz (6) gilt mit (2):

$$|F(X)+R,F(Y)+R| \leqq |F(X),F(Y)| \leqq \Lambda|X,Y|,$$

also ist $F(X)+R$ eine kontrahierende Abbildung auf G. Wie üblich zeigt man damit $|Y_\nu,Y_\mu| \to 0$ für $\nu,\mu \to \infty$. Wegen der metrischen Vollständigkeit von G gibt es dann ein Element $\hat{X} \in G$ mit $Y_\nu \to \hat{X}$. Aus der Stetigkeit von F und aus $Y_{\nu+1} = F(Y_\nu)+R$ folgt dann für $\nu \to \infty$, daß $\hat{X}$ Lösung von (1) ist. Die Eindeutigkeit ist wegen der Regularität von $I - \Lambda$ und wegen $(I-\Lambda)^{-1} \geqq O$ trivial. ■

Für die beiden (ebenfalls wohlbekannten) nächsten Sätze werde vorausgesetzt, daß die betrachtete Abbildung F zusätzlich zu (2) noch <u>inklusionsisoton</u> sein möge, d.h. daß für alle $X,Y \in G$ gelte

$$X \subseteq Y \Rightarrow F(X) \subseteq F(Y). \tag{12}$$

Ferner bestehe zwischen der Inklusionsrelation $\subseteq$ und der Pseudometrik $|\cdot,\cdot|$ die folgende Verträglichkeitsbedingung, die man als Stetigkeit von $\subseteq$ bezüglich metrischer Konvergenz charakterisieren kann:

Aus $X_\nu, X \in G,\ X_\nu \subseteq X_{\nu+1},\ \lim_{\nu\to\infty} X_\nu \to X$

folge $X_\nu \subseteq X$ für alle $\nu = 0,1,\ldots$. (13)

Satz 2: Die Funktion $F: G \to G$ genüge den Bedingungen (2) mit (3) und (12), ferner gelte (13). Dann hängt die (eindeutig bestimmte) Lösung $\hat{X} = \hat{X}(R)$ des Fixmengenproblems (1) isoton von R ab, d.h. es gilt

$$R_o \subseteq R_1 \Rightarrow \hat{X}(R_o) \subseteq \hat{X}(R_1)$$

für alle Elemente $R_o, R_1 \in G$.

Beweis: Es gelte $R_o, R_1 \in G$ mit $R_o \subseteq R_1$ und es sei $\hat{X}(R_o)$ die nach Satz 1 eindeutig bestimmte Lösung von (1) für R_o statt R. Man definiert $Y_o := \hat{X}(R_o)$, $Y_{\nu+1} := F(Y_\nu)+R_1$ für $\nu = 0,1,\ldots$. Damit gilt $Y_o = \hat{X}(R_o) = F(\hat{X}(R_o))+R_o = F(Y_o)+R_o \subseteq F(Y_o)+R_1 = Y_1$. Wegen der Inklusionsisotonie (12) von F folgt daraus $Y_\nu \subseteq Y_{\nu+1}$ für $\nu = 0,1,\ldots$. Nach Satz 1, ii) existiert ein $\hat{X}(R_1) = F(\hat{X}(R_1))+R_1 \in G$ derart, daß $Y_\nu \to \hat{X}(R_1)$ gilt für $\nu \to \infty$. Wegen $\hat{X}(R_o) = Y_o \subseteq Y_\nu \subseteq Y_{\nu+1}$ und wegen der Stetigkeit der Inklusion folgt daraus mit (13) auch $\hat{X}(R_o) \subseteq \hat{X}(R_1)$, was zu beweisen war. ∎

Satz 3: Die Funktion $F: G \to G$ genüge den Bedingungen (2) mit (3) und (12), ferner gelte (13). Es sei $H: G \to G$ eine Abbildung, die derselben Lipschitzbedingung (2) wie F genüge und für die gilt

$$H(X) \subseteq F(X) \quad \text{für alle } X \in G. \tag{14}$$

Dann hat das Fixmengenproblem

$$X = H(X) + R \tag{15}$$

für alle $R \in G$ eine eindeutig bestimmte Lösung $\hat{Z} = \hat{Z}(R) \in G$ und es gilt

$$\hat{Z}(R) \subseteq \hat{X}(R)$$

für alle Elemente $R \in G$ und die eindeutig bestimmte Lösung $\hat{X}(R)$ von (1).

Bemerkungen: 1) Man kann das Ergebnis dieses Satzes 3 verbal auch so formulieren: Die Lösung $\hat{X}$ von (1) hängt isoton ab nicht nur von R, sondern auch von der Abbildung F.

2) Man beachte, daß die Funktion H in Satz 3 <u>nicht</u> notwendigerweise inklusionsisoton zu sein braucht.

<u>Beweis</u>: Es sei $R \in G$ und es sei $\hat{Z}(R) \in G$ die nach Satz 1 eindeutig bestimmte Lösung von (15). Man definiert $Y_o := \hat{Z}(R)$, $Y_{\nu+1} := F(Y_\nu)+R$ für $\nu = 0,1,\ldots$. Nach (14) gilt dann $Y_o = \hat{Z}(R) = H(\hat{Z}(R))+R = H(Y_o)+R \subseteq F(Y_o)+R = Y_1$. Wegen der Inklusionsisotonie (12) von F folgt daraus $Y_\nu \subseteq Y_{\nu+1}$ für $\nu = 0,1,\ldots$. Nach Satz 1, ii) existiert ein $\hat{X}(R) = F(\hat{X}(R))+R \in G$ derart, daß $Y_\nu \to \hat{X}(R)$ gilt für $\nu \to \infty$ und $\hat{X}(R)$ ist Lösung von (1). Wegen $\hat{Z}(R) = Y_o \subseteq Y_\nu \subseteq Y_{\nu+1}$ und wegen der Stetigkeit der Inklusion folgt daraus mit (13) auch $\hat{Z}(R) \subseteq \hat{X}(R)$, was zu beweisen war. ∎

Es sei $S_G \subseteq S$ die von G auf S induzierte Menge

$$S_G := \{a \in S \mid \{a\} \in G\}.$$

Offensichtlich ist S_G mit G abgeschlossen bezüglich der arithmetischen Operation + in S. Man definiert weiter $|\cdot,\cdot|_S$: $S_G \times S_G \to \mathbb{R}^n_+$ durch

$$|a,b|_S := |\{a\},\{b\}| \qquad \text{für } a,b \in S_G.$$

Diese Abbildung ist mit $|\cdot,\cdot|$ eine translations-sub-invariante Pseudometrik auf S_G und S_G ist vollständig bezüglich $|\cdot,\cdot|_S$. Damit läßt sich der Satz 3 über Fix<u>mengen</u> unmittelbar übertragen auf Fix<u>punkt</u>probleme auf S_G durch die

<u>Folgerung aus Satz 3</u>: Die Voraussetzungen von Satz 3 seien erfüllt. Die Funktion f: $S_G \to S_G$ besitze die Eigenschaft

$$f(x) \in F(X) \qquad \text{für } x \in X \in G.$$

Ferner genüge f auf S_G einer Lipschitzbedingung

$$|f(x),f(y)|_S \leqq \Lambda |x,y|_S \qquad \text{für } x,y \in S_G$$

mit der Lipschitzkonstanten $\Lambda \in \mathbb{R}^{n,n}_+$ von Definition (2). Dann hat das Fixpunktproblem

$$x = f(x) + r \tag{16}$$

für beliebige Elemente $r \in S_G$ in S_G genau eine Lösung $\hat{x} = \hat{x}(r)$ und es gilt

$$\hat{x}(r) \in \hat{X}(R)$$

für die Lösung $\hat{X}(R)$ jeder der Aufgaben (1), für die

$$r \in R$$

gilt.

<u>Bemerkung</u>: Wegen (3) läßt sich der Fixpunkt $\hat{x}$ von (16) durch das Punkt-Iterationsverfahren entsprechend zu (11) bestimmen.

Der <u>Beweis</u> der Folgerung geht analog zum Beweis von Satz 3.

Für den folgenden Satz 4 werde noch die Teilmenge $G_o \subseteq G$ eingeführt durch

$$G_o := \{R \in G \mid \text{diam}(A+R) > \text{diam}\, A \quad \text{für alle } A \in G\}. \tag{17}$$

Wegen (7) enthält jedes Element $R \in G_o$ mindestens zwei Punkte $r_o, r_1 \in R$ mit $r_o \neq r_1$. Es gilt weiter

$$\text{zu jedem } r \in R \in G_o \text{ gilt auch } \{r\} \subseteq R. \tag{18}$$

<u>*Satz 4*</u>: *Die Teilmenge $G_o \subseteq G$ nach (17) sei nicht leer. Die Funktion $F\colon G \to G$ genüge einer Lipschitzbedingung (2), ferner gelte mit der darin verwendeten Lipschitzkonstanten $\Lambda \in \mathbb{R}_+^{n,n}$ die Abschätzung*

$$\text{diam}\, F(X) \geqq \Lambda\, \text{diam}\, X \qquad \text{für } X \in G. \tag{4}$$

Dann sind die folgenden 6 Aussagen äquivalent:

i) Die Fixmengengleichung

$$X = F(X) + R \tag{1}$$

hat für alle Elemente $R \in G$ je eine eindeutig bestimmte Lösung $\hat{X} = \hat{X}(R) \in G$.

ii) Die (spezielle) Fixmengengleichung

$$X = F(X) + R_o$$

mit festem $R_o \in G_o$ besitzt mindestens eine Lösung $X_o \in G$.

iii) Es gibt Elemente $X_1 \in G$, $R_1 \in G_o$ derart, daß die Inklusion

$$X_1 \supseteq F(X_1) + R_1 \tag{19}$$

gilt.

iv) *Es gibt Elemente $X_2, R_2 \in G$ derart, daß die strikte Inklusion*

$$X_2 \dot{\supset} F(X_2) + R_2 \tag{20}$$

gilt.

v) *Für den Spektralradius $\rho(\Lambda)$ der Lipschitzkonstanten Λ in (2) gilt*

$$\rho(\Lambda) < 1. \tag{3}$$

vi) *Zu $R \in G$ sei die Folge $\{Y_\nu\}$ mit $Y_\nu \in G$ durch das Iterationsverfahren*

$$\left.\begin{aligned} &Y_0 \in G, \\ &Y_{\nu+1} := F(Y_\nu) + R \qquad \text{für } \nu = 0,1,\ldots \end{aligned}\right\} \tag{11}$$

definiert. Dann gilt für alle Anfangswerte $Y_0 \in G$ stets

$$\lim_{\nu\to\infty} Y_\nu = \hat{X}.$$

Dabei ist $\hat{X} \in G$ die in G eindeutig bestimmte Fixmengenlösung zu (1).

Beweis: Der Beweis der behaupteten Äquivalenzen wird geführt, indem die Implikationskette i) ⇒ ii) ⇒ ... ⇒ vi) ⇒ i) gezeigt wird.

i) ⇒ii): Trivial wegen $G_0 \subseteq G$ mit $X_0 := \hat{X}(R)$.

ii) ⇒iii): Trivial mit $R_1 := R_0$, $X_1 := X_0$.

iii)⇒iv): Man setzt $X_2 := X_1$, wählt ein beliebiges $r \in R_1$, für das nach (18) $\{r\} \subseteq R_1$ gilt und setzt $R_2 := \{r\} \subseteq R_1$. Damit gilt nach (19) die Inklusionskette $X_2 = X_1 \supseteq F(X_1)+R_1 = F(X_2)+R_1 \supseteq F(X_2)+R_2$. Wegen $R_1 \in G_0$, $R_2 \subseteq R_1$, (9) und (7) folgt daraus die Ungleichungskette

$$\text{diam } X_2 \geq \text{diam } (F(X_2)+R_1) > \text{diam } F(X_2) = \text{diam } (F(X_2)+R_2),$$

also gilt (20) nach (10).

iv) ⇒ v): Aus (20) folgt wegen (8), (7) und (4) diam X_2 > diam $(F(X_2)+R_2)$ $\geqq$ diam $F(X_2) \geqq \Lambda$ diam X_2, oder mit ξ := diam X_2 die Ungleichung

$$\xi > \Lambda\xi .$$

Nach einem bekannten Satz über nichtnegative Matrizen (siehe etwa Varga [6], Seite 87) folgt daraus (3).

v) ⇒ vi) ⇒ i): Diese Implikationen sind der Inhalt von Satz 1. ∎

4. Anwendungen.

4.1. Beispiel 1

Der lineare Raum S über dem Konstantenkörper $\mathbb{R}$ sei in dem folgenden schwachen Sinne ein n-dimensionaler Raum: Auf S gebe es eine translations-sub-invariante n-dimensionale Pseudometrik $|\cdot,\cdot|_S$: $S \times S \to \mathbb{R}^n_+$ und S sei bezüglich dieser Metrik vollständig. Wie üblich wird mit dem Nullelement $\vartheta \in S$ eine Pseudonorm $|\cdot|_S$: $S \to \mathbb{R}^n_+$ (mit dem Namen "Betrag") auf S eingeführt durch die Setzung

$$|a|_S := |a,\vartheta|_S \qquad \text{für alle } a \in S.$$

Als Grundmenge $G \subseteq \mathbb{P}(S)$ wird bei diesem Beispiel 1 die Menge aller kompakten Teilmengen aus $\mathbb{P}(S)$ betrachtet. Durch (5) wird die arithmetische Operation + von S auf G übertragen; offenbar ist G abgeschlossen bezüglich dieser Verknüpfung. Als Pseudometrik $|\cdot,\cdot|$: $G \times G \to \mathbb{R}^n_+$ auf G wird die Hausdorff-Metrik verwendet, sie ist definiert durch

$$A,B := \max\left(\sup_{a\in A}\ \inf_{b\in B} |a,b|_S,\ \sup_{b'\in B}\ \inf_{a'\in A} |a',b'|_S\right)$$

für alle A,B ∈ G. Man zeigt leicht die folgenden

Eigenschaften:

i) Der Raum G ist metrisch vollständig (siehe etwa Hausdorff [2], S. 150 VI).

ii) Ist $|\cdot,\cdot|_S$ translationsinvariant oder translations-subinvariant, dann ist $|\cdot,\cdot|$ in beiden Fällen translations-subinvariant.

iii) Die Verträglichkeitsbedingung (13) ist erfüllt.

iv) Die Pseudometrik $|\cdot,\cdot|$ ist inklusionsisoton.

Es sei $f: S \to S$ und f genüge auf S einer Lipschitzbedingung

$$|f(x),f(y)|_S \leqq \Lambda |x,y|_S \qquad \text{für } x,y \in S \tag{21}$$

mit einer Lipschitzkonstanten $\Lambda \in \mathbb{R}_+^{n,n}$. Man definiert die Funktion $F: G \to \mathbb{P}(S)$ durch

$$F(X) := \{f(x) \mid x \in X\} \qquad \text{für } X \in G. \tag{22}$$

Wegen (21) ist f auf S stetig, damit ist mit X auch die Menge $F(X)$ kompakt, also wird durch (22) sogar eine Abbildung $F: G \to G$ definiert. Offensichtlich genügt auch F auf G wieder einer Lipschitzbedingung

$$|F(X),F(Y)| \leqq \Lambda |X,Y| \qquad \text{für } X,Y \in G \tag{2}$$

mit derselben Lipschitzkonstanten Λ wie in (21). Ferner ist F auch inklusionsisoton. Damit lassen sich die Sätze 1 bis 3 auf die hier betrachteten speziellen Räume S und G und die Funktion F aus (22) anwenden.

Um auch Satz 4 verwenden zu können, ist zunächst zu prüfen, ob die Teilmenge

$$G_o := \{R \in G \mid \operatorname{diam}(A+R) > \operatorname{diam} A \text{ für } A \in G\} \subseteq G$$

nicht leer ist. Es gilt das

<u>Lemma:</u> *In dem linearen Raum S gebe es eine Basis $s_1, s_2, \ldots, s_n$. Dann ist G_o nicht leer.*

<u>Beweis:</u> Man setzt mit einer Norm $\|\cdot\|$ auf $\mathbb{R}^n$:

$R_o := \left\{ \sum_{\nu=1}^{n} \alpha_\nu s_\nu \mid \alpha_\nu \in \mathbb{R}, \ \left\| \left| \sum_{\nu=1}^{n} \alpha_\nu s_\nu \right|_S \right\| \leqq 1 \right\}$. Es gilt $\operatorname{diam} R_o > 0$. Da R_o kompakt ist, gilt weiter $R_o \in G$.

Es sei $A \in G$ beliebig gewählt. Ist $\operatorname{diam} A = 0$, also $A = \{a\}$, dann gilt nach (7):

$$\operatorname{diam}(A+R_o) = \operatorname{diam} R_o > 0 = \operatorname{diam} A,$$

also gilt dann $R_o \in G_o$.

Ist andererseits diam $A \neq 0$, dann gibt es zwei Elemente $a,a' \in A$ mit $a \neq a'$ und diam $A = |a-a'|_S$. Da die Elemente $\{s_\nu\}$ eine Basis in S bilden, gibt es Konstante $\beta_\nu \in \mathbb{R}$ derart, daß gilt

$$a-a' = b := \sum_{\nu=1}^{n} \beta_\nu s_\nu .$$

Wählt man $b_o := b/\|\,|b|\,\|$, dann ist $b_o \in R_o$. Mithin gilt wegen $a,a' \in A$ und $\theta, b_o \in R_o$:

$$\begin{aligned}\operatorname{diam}(A+R_o) &:= |A+R_o - (A+R_o), \theta|_G \\ &= \sup_{\substack{a'',a''' \in A \\ r,r' \in R_o}} |a''-a'''+r-r'|_S \\ &\geqq |a-a'+b_o-\theta|_S \\ &= \left|(a-a')\left(1+\frac{1}{\|\,|a-a'|\,\|}\right)\right|_S \\ &= \left(1+\frac{1}{\|\,|a-a'|\,\|}\right)|a-a'|_S \\ &> |a-a'|_S = \operatorname{diam} A,\end{aligned}$$

also ist auch in diesem Falle $R_o \in G_o$. ∎

Um nunmehr Satz 4 anwenden zu können, wäre noch zusätzlich die Abschätzung

$$\operatorname{diam} F(X) \geqq \Lambda \operatorname{diam} X \tag{4}$$

für die Funktion F und alle $X \in G$ erforderlich. Aus (21) findet man jedoch mit (22) und (2) die "gegenteilige" Ungleichung

$$\operatorname{diam} F(X) \leqq \Lambda \operatorname{diam} X.$$

Allein für Funktionen $f\colon S \to S$ mit der Eigenschaft

$$|f(x),f(y)|_S = \Lambda |x,y|_S \quad \text{für } x,y \in S \tag{23}$$

gilt (4) (mit dem =-Zeichen).

Auf die Klasse der durch (22) definierten Funktionen ist daher Satz 4 nur dann anwendbar, wenn die "Ausgangsfunktion" f in dem durch (23) definierten Sinne "linear" ist. Man beachte, daß (23) nicht nur eine Forderung an f ist, sondern auch an die Räume S und G und an die Metrik $|\cdot,\circ|_S$.

Nach diesem Beispiel 1 könnte zunächst der Eindruck entstehen, daß der Anwendungsbereich des Satzes 4 nur sehr eingeschränkt sei. Geht man jedoch von der "natürlichen" Definition (22) für die Funktion F ab, dann findet man leicht große Funktionenklassen, für die die Eigenschaft (4) zutrifft und auf die daher Satz 4 anwendbar ist. Ein spezieller derartiger Fall ist dargestellt in dem folgenden

4.2 Beispiel 2

Wie in Beispiel 1 sei der lineare Raum S über dem Konstantenkörper $\mathbb{R}$ vollständig bezüglich einer translations-sub-invarianten n-dimensionalen Pseudometrik $|\cdot,\cdot|_S\colon S\times S \to \mathbb{R}^n_+$. Ist $a \in S$ und $\alpha \in \mathbb{R}^n_+$, dann wird durch

$$A = \langle a,\alpha\rangle := \{x \in S \mid |x,a|_S \leqq \alpha\}$$

eine n-dimensionale Pseudokugel mit dem Mittelpunkt a und dem Radius α definiert.
Es sei

$$G := \{\langle a,\alpha\rangle \mid a \in S,\ \alpha \in \mathbb{R}^n_+\} \subseteq \mathbb{P}(S)$$

die Menge aller dieser Pseudokugeln. Ist $A = \langle a,\alpha\rangle$, $B = \langle b,\beta\rangle \in G$, dann gilt

$$A = B \Leftrightarrow a = b \wedge \alpha = \beta$$

und

$$A \subseteq B \Leftrightarrow |a,b|_S \leqq \beta - \alpha\ .$$

Für die Erweiterung der arithmetischen Operation + von S auf G gemäß (5) findet man entsprechend

$$A + B = \langle a+b,\ \alpha+\beta\rangle$$

und (24)

$$A - B = \langle a-b,\ \alpha+\beta\rangle.$$

Insbesondere gilt damit

$$A - A = \langle \vartheta, 2\alpha\rangle$$

und

$$\Theta = \langle \vartheta, 0\rangle.$$

Die Menge G ist offenbar abgeschlossen bezüglich der Verknüpfung + nach (24).

Man setzt weiter

$$|A,B| := |a,b|_S + |\alpha - \beta| \tag{25}$$

mit dem üblichen komponentenweisen Vektorbetrag $|\cdot|: \mathbb{R}^n \to \mathbb{R}^n_+$. Damit ist dann auf G eine translations-sub-invariante und inklusionsisotone Pseudometrik erklärt und G ist metrisch vollständig bezüglich dieser Pseudometrik. Man zeigt leicht, daß die Verträglichkeitsbedingung (13) auch hier gilt. Ferner ist

$$\operatorname{diam} A := |A-A,\Theta| = 2\alpha, \tag{26}$$

also besitzt der Durchmesser "diam" auch bei den hier betrachteten Pseudokugeln die übliche Eigenschaft, das Doppelte des Radius zu sein. Dieser Durchmesser besitzt die oben genannten Eigenschaften (7) bis (9), dabei ist (8) sogar noch verschärft zu

$$\operatorname{diam}(A+B) = \operatorname{diam} A + \operatorname{diam} B.$$

Damit ist

$$\operatorname{diam}(A+R) > \operatorname{diam} A$$

genau dann, wenn diam R > 0 ist. Also ist die Menge G_0 nach (17) bei diesem Beispiel 2 genau die Menge der "dicken" Pseudokugeln, d.h. es gilt

$$G_0 = \{R \in G \mid \operatorname{diam} R > 0\}.$$

Es sei $f\colon S \to S$ eine Funktion, die wie im Beispiel 1 einer Lipschitzbedingung (21) mit $\Lambda \in \mathbb{R}_+^{n,n}$ genüge. Man definiert die Funktion $F\colon G \to G$ gemäß

$$F(X) := \langle f(x), \Lambda\xi \rangle \qquad \text{für } X = \langle x,\xi \rangle \in G. \tag{27}$$

Mit (21) und (25) findet man sofort, daß die Funktion F auf G der Lipschitzbedingung (2) genügt. Weiterhin ist nach (26) und (27)

$$\operatorname{diam} F(X) = 2\Lambda\xi = \Lambda \operatorname{diam} X$$

für alle $X = \langle x,\xi \rangle \in G$, also ist auch (4) (mit dem =-Zeichen) befriedigt. Damit lassen sich die Sätze 1 bis 4 ohne jede Einschränkung auf dieses zweite Beispiel anwenden.

Man zeigt leicht, daß dasselbe auch richtig ist für passend definierte Funktionen auf Intervallvektoren, wenn der lineare Raum S zusätzlich noch halbgeordnet ist.

5. Erweiterungen.

5.1. Der Durchmesser "diam" wird nur in Satz 4 benutzt, nicht dagegen für die Sätze 1 bis 3. Analysiert man den Beweis des Satzes 4 so sieht man, daß darin gar nicht benötigt wird, daß die Abbildung "diam" gerade der Durchmesser der betrachteten Mengen ist. Man kann daher in Satz 4 den Durchmesser durch jedes andere n-dimensionale Pseudofunktional $\operatorname{diam}\colon G \to \mathbb{R}_+^n$ ersetzten, falls dieses nur die Eigenschaften (7), (8) und (9) besitzt.

5.2. Anstelle einer speziellen Pseudometrik $|\cdot,\cdot|\colon G\times G \to \mathbb{R}_+^n$ kann allgemeiner eine Abbildung $|\cdot,\cdot|\colon G\times G \to H$ zugelassen werden, wobei H ein passender, z.B. linearer halbgeordneter Raum ist. Entsprechend ist dann auch das Pseudofunktional diam durch eine Abbildung $\operatorname{diam}\colon G \to H$ zu ersetzen. An die Stelle der Matrix $\Lambda \in \mathbb{R}_+^{n,n}$ tritt in der Lipschitzbedingung (2) und in der Durchmesserbedingung (4) dann ein (z.B. linearer) Operator $\Lambda\colon H \to H$. In dieser Formulierung kann dann auch noch auf die Einschränkung $n < \infty$ der endlichen Dimension von H verzichtet werden.

5.3. Genügt die Funktion F(X) einer Lipschitzbedingung (2), so gilt - wegen der Translations-sub-Invarianz (6) - dasselbe auch für die rechte Seite F(X)+R der Gleichung (1). Wegen (8) ist diam (F(X)+R) $\geq$ diam F(X), d.h. mit F(X) befriedigt auch F(X)+R die Durchmesserbedingung (4). Man kann daher fragen, warum in der vorliegenden Arbeit nicht "die Menge R in die Funktion F(X) mithinein genommen wurde", d.h. warum nicht die Gleichung (1) durch die Fixmengengleichung X = F(X) ersetzt wurde? Die ersten drei Sätze ließen sich unmittelbar übertragen. Beim Satz 4 wären die Aussagen ii) und iii) passend zu modifizieren. Insbesondere jedoch wäre in der Aussage i) der Satz: "das Problem (1) hat für alle Elemente R $\in$ G je eine eindeutig bestimmte Lösung $\hat{X} \in G$" zu ersetzen durch: "Das Fixmengenproblem X = F(X) hat für die Menge aller Abbildungen F: G $\rightarrow$ G, die den Bedingungen (2) und (4) genügen, je eine eindeutig bestimmte Lösung $\hat{X} \in G$".

Die so angedeutete Änderung hätte den Vorteil, daß dann in den Aussagen der Sätze 1 bis 4 keine Addition mehr auftreten würde. Damit könnte die oben benutzte Grundmenge G, auf der neben der Inklusion und der Pseudometrik eine Addition + erklärt ist, ersetzt werden durch einen beliebigen halbgeordneten und mit einer Pseudometrik versehenen Grundbereich. Insbesondere könnte man dann auch auf die Elementmenge S verzichten.

Zwei Nachteile dieses Vorgehens wären unter anderen, daß die Anschaulichkeit der Definition der Teilmenge $G_o \subseteq G$ verloren ginge und daß die Anwendbarkeit des Satzes 4 - etwa in der Intervall-Mathematik - nicht mehr so unmittelbar ersichtlich wäre. Aus diesen Gründen wurde hier auf die angedeutete abstraktere Formulierung verzichtet.

Literatur

[1] Collatz, L.: Funktionalanalysis und Numerische Mathematik. Springer 1964

[2] Hausdorff, F.: Mengenlehre. Walter de Gruyter (Göschen) 1927

[3] Kurepa, G.: Tableaux ramifiés d' ensembles, Espaces pseudodistanciés. C.R. Acad. Sc. Paris 198 (1934), 1563-1565

[4] Davis, Ph.J., Najfeld, I. and Vitale,R.A.: Minkowski Iteration of Sets. Linear Algebra and its Applications 29 (1980), 259-291

[5] Nickel, K.: Die Auflösbarkeit linearer Kreisscheiben- und Intervall-Gleichungssysteme. Freiburg Intervall-Berichte 81/3 (1981), 11-48

[6] Varga, R.S.: Matrix Iterative Analysis. Englewood Cliffs: Prentice Hall Inc. (1962)

ON THE CONVERGENCE

OF A CLASS OF NEWTON-LIKE METHODS

F.-A.Potra
Department of Mathematics
The National Institute
for Scientific and Technical Creation
Bd.Pacii 220, 79622 Bucharest, Romania

1. Introduction

In a previous paper [13] we have studied a class of iterative procedures of the form

(1) $$Fx_n+\delta F(x_{p_n},x_{q_n})(x_{n+1}-x_n)=0\ , \qquad n=0,1,2,\dots$$

where F was a nonlinear operator between two Banach spaces, δF a strongly consistent approximation of F', and $(p_n)_{n\geq 0}$, $(q_n)_{n\geq 0}$ two sequences of integers satisfying the condition

(2) $$-1=q_o\leq q_n\leq n\ , \quad 0=p_o\leq p_n\leq n\ , \qquad q_n\leq p_n\ .$$

This iterative procedure reduces to Newton's method for $p_n=q_n=n$ and to the secant method for $p_n=n$, $q_n=n-1$. The choice

(3) $$p_{km+j}=q_{km+j}=km\ , \qquad j=0,1,\dots,m-1;\ k=0,1,2,..$$

where m is a fixed positive integer leads to an iterative procedure investigated in [2], [3], [15], [20], [22], [24]. If

(4) $$p_{km+j}=km\ ,\ q_{km+j}=km-1\ , \qquad j=0,1,\dots,m-1;\ k=0,1,2,\dots$$

then (1) reduces to the so called multi-step secant method (see [5], [9], [16], [20]).

It can be proved that the sequence $(z_k)_{k\geq 0}$, with, $z_k=x_{km}$ has the order of convergence equal to m+1 in case of choice (3), and $\frac{m+\sqrt{m^2+4}}{2}$ in case of choice (4). The parameter m can be chosen according to the dimension of the space as to maximize the efficiency of the procedure (see [20]).

Usually $\delta F(x,y)\neq\delta F(y,x)$ so that together with (1) we may also consider the iterative procedure

(5) $$Fx_n+ F(x_{q_n},x_{p_n})(x_{n+1}-x_n)=0\ , \qquad n=0,1,2,\dots\ .$$

More generally we can set

(6) $\quad Fx_n+A_n(x_{n+1}-x_n)=0 \;, \qquad n=0,1,2,\ldots$

where for each n, A_n can be taken either $\delta F(x_{p_n},x_{q_n})$ or $\delta F(x_{q_n},x_{p_n})$ i.e:

(7) $\quad A_n \varepsilon \{\delta F(x_{p_n},x_{q_n}) \;,\; \delta F(x_{q_n},x_{p_n})\} \;, \qquad n=0,1,2,\ldots$

In the first part of the present paper we shall make a semi-local analysis for this iterative procedure obtaining a slight generalization of a result contained in [13]. This will improve also a result of J.E.Dennis [5].

In the second part of the paper we shall investigate the monotonicity properties of an iterative procedure of type (6), (7) generalizing some results of [11], [17], [19], [23], [25].

2. Notation

If X and Y are two linear spaces (L-spaces) then we denote by L(X,Y) the set of all linear operators from X into Y. If X and Y are topological linear spaces (TL-spaces) then B(X,Y) denotes the set of bounded linear operators from X into Y. If X and Y are normed spaces (N-spaces) then the space B(X,Y) is endowed with the operator norm. All the norms will be denoted by the symbol $|\cdot|$.

A subset K of an L-space X is called cone if $K+K\subset K$ and $\alpha K\subset K$ for $\alpha>0$. The cone K is called proper if $K\cap\{-K\}=\{0\}$. If K is a proper cone of X then the relation "$\leq$", defined by $x\leq y \Leftrightarrow y-x\varepsilon K$ is a partial ordering in X. An L-space X endowed with such a relation is called a partially ordered linear space (POL-space). Two elements x and y of X are called comparable if either $x\leq y$ or $y\leq x$ holds. If x and y are comparable then we denote by $x\wedge y$ (resp. $x\vee y$) the minimum (resp.maximum) of x and y. If $a\leq b$ then we denote by $[a,b]$ the set $\{x\varepsilon X;\; a\leq x\leq b\}$. If u, v are two comparable points of X then we denote by $\langle u,v\rangle$ either the set $[u,v]$, if $u\leq v$, or the set $[v,u]$, if $u\geq v$.

A TL-space partially ordered by a closed proper cone is called a partially ordered topological linear space (POTL-space). A POTL-space X is called normal if, given a local basis $\mathcal{U}$ for the topology, there exists a positive number η so that if $0\leq z\varepsilon U\varepsilon\mathcal{U}$ then $[0,z]\subset\eta U$. A POTL-space is called regular if every order bounded increasing sequence has a limit. We note that any regular partially ordered Banach space is normal but the reverse is not true. For example the space C[0,1] with the natural partial ordering is normal but not regular. All finite dimensional POTL-spaces are both normal and regular.

Let X and Y be two POL-spaces and G an operator from X into Y.

G is called: nonnegative if $Gz\geq 0$ for all $z\geq 0$; inverse nonnegative if $Gz\geq 0$ implies $z\geq 0$; isotone if $z_1\leq z_2$ implies $Gz_1\leq Gz_2$, antitone if $z_1\leq z_2$ implies $Gz_1\geq Gz_2$. If G is nonnegative we write $G\geq 0$. If G and H are two operators such that $G-H\geq 0$ then we write $H\leq G$. The space L(X,Y) with the relation "≤" defined above becomes a POL-space. Let T be an operator belonging to L(X,Y). An operator $S\in L(Y,X)$ is called a left (resp.right) subinverse of T if $ST\leq I$ (resp.$TS\leq I$) where I denotes the identity operator in X (resp. in Y). S is called a subinverse of T if it is a left as well as a right subinverse of T.

3. Semi-local convergence

3.1. Theorem. Let F be a nonlinear operator defined on a convex subset D of a Banach space X, with values in a Banach space Y, and let x_o,x_{-1} be two points from the interior $\overset{o}{D}$ of D satisfying the inequality

(8) $$|x_o-x_{-1}|\leq c .$$

Suppose F is Fréchet differentiable on $\overset{o}{D}$ and there exists a mapping $\delta F:\overset{o}{D}\times\overset{o}{D}\to B(X,Y)$ such that the linear operator A_o , where A_o is either $\delta F(x_o,x_{-1})$ or $\delta F(x_{-1},x_o)$, is invertible, its inverse T_o is continuous and:

(9) $$|T_oFx_o|\leq b,$$

(10) $$|T_o(\delta F(x,y)-F'(z))|\leq a(|x-z|+|y-z|), \qquad x,y,z\in\overset{o}{D} .$$

Let $(p_n)_{n\geq 0}$ and $(q_n)_{n\geq 0}$ be two sequences of integers satisfying condition (2).

If the constants a, b, c introduced above satisfy the inequality

(11) $$ac+2(ab)^{1/2}\leq 1$$

and if the set $D_c=\{x\in D;\ f \text{ is continuous at } x\}$ contains the closed ball U with center $x_1=x_o-T_oFx_o$ and radius $r_1=\frac{1}{2a}\{1-a(2b+c)-[(1-ac)^2-4ab]^{1/2}\}$, then the iterative procedure described by (6), (7) is well defined (i.e. for each n there exists $x_{n+1}\in\overset{o}{D}$ satisfying (6)), the sequence $(x_n)_{n\geq 1}$ produced by it converges to a root $x^*\in U$ of the equation $Fx=0$ and the following estimates hold:

(12) $$|x_n-x^*|\leq t_o-|x_n-x_o|-[(t_o-|x_n-x_o|)^2-(|x_n-x_{p_{n-1}}|+ +|x_{n-1}-x_{p_{n-1}}|+|x_{p_{n-1}}-x_{q_{n-1}}|)|x_n-x_{n-1}|]^{1/2}\leq t_n-d ,$$

(13) $$|x_n-x^*|\geq[(t_o-2^{-1}(|x_{p_n}-x_{q_n}|+|x_{p_n}-x_o|+|x_{q_n}-x_o|)- -|x_n-x_{p_n}|)^2+(2t_o-|x_{p_n}-x_o|-|x_{q_n}-x_o|)|x_n-x_{n+1}|]^{1/2}-$$

$$-t_o+2^{-1}(|x_{p_n}-x_{q_n}|+|x_{p_n}-x_o|+|x_{q_n}-x_o|)+|x_n-x_{p_n}|$$

where

(14) $$d=\frac{1}{2a}[(1-ac)^2-4ab]^{1/2},$$

(15) $$t_{-1}=\frac{1+ac}{2a},\quad t_o=\frac{1-ac}{2a},\quad t_{n+1}=t_n-\frac{t_n^2-d^2}{t_{p_n}+t_{q_n}},\qquad n=0,1,2,\ldots .$$

Proof. We shall follow closely the proof of Theorem 1 of [13]. First we observe that the linear operator $P=\delta F(u,v)$ is invertible for all $u,v\in\overset{o}{D}$ with

(16) $$|u-x_o|+|v-x_o|<2t_o .$$

Indeed from (10) it follows that

$$|I-T_oP|=|T_o(A_o-P)|\le|T_o(P-F'(x_o))|+|T_o(F'(x_o)-A_o)|\le$$

$$\le a(|u-x_o|+|v-x_o|+|x_o-x_{-1}|)<1$$

so that, according to Banach's lemma, P is invertible and

(17) $$|(T_oP)^{-1}|\le[1-a(|u-x_o|+|v-x_o|+c)]^{-1} .$$

Let us note that condition (10) implies the following Lipschitz condition for F':

(18) $$|T_o(F'(u)-F'(v))|\le 2a|u-v|,\qquad u,v\in\overset{o}{D} .$$

Using the integral representation

(19) $$Fx-Fy=[\int_0^1 F'(y+t(x-y))dt](x-y)$$

we deduce that

(20) $$|T_o[Fx-Fy-F'(u)(x-y)]|\le a(|x-u|+|y-u|)|x-y|$$

for all $x,y,u\in\overset{o}{D}$.

Finally from (10) and (20) we have

(21) $$|T_o[Fx-Fy-\delta F(u,v)(x-y)]|\le a(|x-u|+|y-u|+|u-v|)|x-v|$$

for all $x,y,u,v\in\overset{o}{D}$. By a continuity argument (19), (20) and (21) remain valid if x and/or y belong to D_c.

Using the above inequalities we shall prove that

(22) $$|x_n-x_{n+1}|\le t_n-t_{n+1}$$

for $n=-1,0,1,2,\ldots$

It is easy to see that the sequence $(t_n)_{n\ge -1}$ given by (15) is decreasing and converges to d. If k is a nonnegative integer and if (22) holds for $n\le k-1$ then:

$$|x_o-x_n|\le t_o-t_n<t_o-d\ ,\quad |x_1-x_n|\le t_1-t_n<t_1-d=r_1$$

for $n\le k$. This shows that (16) is satisfied for $u=x_i$ and $v=x_j$ with $i,j\le k$. Thus (22) assures the fact that (6) is well defined.

For $n=-1$ and $n=0$ (22) reduces to $|x_{-1}-x_o|\le c$ and $|x_o-x_1|\le b$ (compare with (8) and (9)). Suppose (22) holds for $n=-1,0,\dots,k$, where $k\ge 0$. Denote $T_n=A_n^{-1}$.

Using (6), (17) and (21) we may write:

$$|x_{k+1}-x_{k+2}|=|T_{k+1}Fx_{k+1}|=|(T_oA_{k+1})^{-1}T_o[Fx_{k+1}-Fx_k-A_k(x_{k+1}-x_k)]|\le$$

$$\le\frac{a(|x_{k+1}-x_{p_k}|+|x_k-x_{p_k}|+|x_{p_k}-x_{q_k}|)}{1-a(|x_{p_{k+1}}-x_o|+|x_{q_{k+1}}-x_o|+c)}|x_k-x_{k+1}|\le$$

$$\le\frac{a(t_{p_k}-t_{k+1}+t_{p_k}-t_k+t_{q_k}-t_{p_k})}{1-a(t_o-t_{p_{k+1}}+t_o-t_{q_{k+1}}+t_{-1}-t_o)}(t_k-t_{k+1})=$$

$$=\frac{t_{p_k}+t_{q_k}-t_{k+1}-t_k}{t_{p_{k+1}}+t_{q_{k+1}}}(t_k-t_{k+1})=t_{k+1}-t_{k+2}\ .$$

We have thus proved that (22) holds for all n. From the completeness of X it follows that the sequence $(x_n)_{n\ge 0}$ converges to a point x^* and that

$$(23)\qquad |x_n-x^*|\le t_n-d\ .$$

From (6), (7) and (22) we obtain the inequality

$$(24)\quad |T_oFx_{k+1}|\le a(|x_{k+1}-x_{p_k}|+|x_k-x_{p_k}|+|x_{p_k}-x_{q_k}|)|x_k-x_{k+1}|$$

wherefrom it follows that $Fx^*=0$.

Let us take now $x=x_n$ and $y=x^*$ in (19) and denote $S=\int_0^1 F'(x^*+t(x_n-x^*))dt$. According to (22) and (23) it follows that

$$|x_n-x_o|+|x^*-x_o|+|x_o-x_{-1}|\le 2|x_n-x_o|+|x_n-x^*|+c<$$

$$<c+2(|x_n-x_o|+|x_n-x^*|)\le 2(t_o-t_n+t_n-d)+c\le 2t_o+c=1/a\ .$$

Using (10) and Banach's lemma, one can prove that S is invertible and

$$(25)\qquad |(T_oS)^{-1}|\le[1-a(2|x_n-x_o|+|x_n-x^*|+c)]^{-1}$$

(see also the proof of inequality (17).) According to (24) and (25) we have

$$|x_n-x^*|=|S^{-1}Fx_n|\le|(T_oS)^{-1}||T_oFx_n|\le$$

$$\leq\frac{a(|x_n-x_{p_{n-1}}|+|x_{n-1}-x_{p_{n-1}}|+|x_{p_{n-1}}-x_{q_{n-1}}|)}{1-a(2|x_n-x_o|+|x_n-x^*|+c)}|x_n-x_{n-1}|$$

and it is easy to see that the above inequality together with the fact $|x_n-x^*|<t_o$ implies the estimate (12).

Using the identity

$$x_{n+1}-x_n=x^*-x_n+(T_oA_n)^{-1}T_o(Fx^*-Fx_n)-A_n(x^*-x_n))$$

and the inequalities (17) and (21) we obtain

$$|x_{n+1}-x_n|\leq\frac{a(2|x_n-x_{p_n}|+|x^*-x_n|+|x_{p_n}-x_{q_n}|)}{1-a(|x_{p_n}-x_o|+|x_{q_n}-x_o|+c)}|x_n-x^*|+|x_n-x^*|.$$

This inequality implies the lower bound (13) ∎

Let us observe that by taking x=y in (10) we deduce that the Fréchet derivative of F satisfies a Lipschitz condition of the form

$$|T_o(F'(y)-F'(z))|\leq 2a|y-z|.$$

Conversely if the above condition is satisfied then (10) is also fulfilled, taking for example $\delta F(x,y)=\int_0^1 F'(tx+(1-t)y)dt$. It follows that for $x_o=x_{-1}$ and c=0 the hypothesis of Theorem 3.1 reduces to the hypothesis of the affine invariant version of the Kantorovich theorem [6]. We obtain thus the following:

3.2. Corollary. Let X, Y be two Banach spaces and D a convex subset of X. Let $F:D\subset X\to Y$ be a nonlinear operator, Fréchet differentiable on $\mathring{D}$. Suppose that, for an $x_o\in\mathring{D}$, $F'(x_o)$ is invertible and its inverse $T_o=[F'(x_o)]^{-1}$ is continuous. If there exist two constants a, b such that

$$4ab\leq 1,\quad |T_oFx_o|\leq b,\quad |T_o(F'(x)-F'(y))|\leq 2a|x-y|,\quad x,y\in\mathring{D},$$

and if $U:=\{x\in X;\ |x-x_o+T_oFx_o|\leq\frac{1}{2a}[1-2ab-(1-4ab)^{1/2}]\}\subset D_c:=\{x\in D;\ F$ is continuous at $x\}$, then Newton's method:

(26) $$Fx_n+F'(x_n)(x_{n+1}-x_n)=0 \qquad n=0,1,2,\ldots$$

is well defined, the sequence $(x_n)_{n\geq 0}$ produced by it converges to a root $x^*\in U$ of the equation Fx=0, and the following estimates hold:

(27) $$|x_n-x\ |\leq t_o-|x_n-x_o|-[(t_o-|x_n-x_o|)^2-|x_n-x_o|^2]^{1/2},$$

(28) $$|x_n-x\ |\geq[(t_o-|x_n-x_o|)^2+2(t_o-|x_n-x_o|)|x_n-x_o|)|x_n-x_{n+1}|]^{1/2}-t_o+|x_n-x_o|,$$

where $t_o=(2a)^{-1}$ ■

It is interesting to note that the error bounds (27) and (28), which follow from (12) and (13) for c=0 and $p_n=q_n=n$, are generally more accurate than the error estimates obtained especially for Newton's method by Gragg and Tapia [7], Miel [10], Potra and Pták [14] (see [12]). We also note that the error bounds obtained in Theorem 3.1 are sharp in the following sense:

3.3. Proposition. If $a>0$, $b\geq 0$, $c\geq 0$ are three constants satisfying inequality (11) then:

(i) There exist a function $F:R\to R$ and two points $x_o, x_{-1}\in R$ verifying the hypothesis of Theorem 3.1 and for which the estimates (12) are attained at each n=1,2,3,... .

(ii) For any given $n\in Z_+$ there exist a function $f_n:R\to R$ and two points $x_o, x_{-1}\in R$ verifying the hypothesis of Theorem 3.1 and for which (13) holds with equality.

Proof. (i) $f(x)=x^2-d^2$, $x_o=t_o$, $x_{-1}=t_{-1}$;

(ii) $f_n(x)=x^2-d^2$ if $x\geq t_n$ and $f_n(x)=-x^2+4t_nx-2t_n^2-d^2$ if $x<t_n$; $x_o=t_o$, $x_{-1}=t-1$ ■

4. Monotonous convergence

In this section supposing that the operator F acts between two POL-spaces we shall use iterative procedures of type (6) in order to obtain monotonically convergent sequences enclosing the roots of the equation Fx=0.

First let us introduce some notation: Let x_o, x_{-1}, y_o be three comparable points belonging to the domain of definition D of the operator F such that

(29) $$x_o\in\langle x_{-1} , y_o\rangle\subset D .$$

We denote by D_1 the set

(30) $$D_1=\{(x,y)\in\langle x_{-1},y_o\rangle^2;\ x \text{ and } y \text{ are comparable}\}$$

and we consider a mapping A defined on D_1 and taking values linear operators. In the statement of the next theorems we shall use the following set of hypotheses and conclusions:

(31) $$x_o\leq y_o , \quad Fx_o\geq 0\geq Fy_o ,$$

(32) $$Fy-Fx\geq A(u,v)(y-x) , \quad u\vee v\leq x\leq v\leq v_o .$$

(H_1)

(33)	$x_o \geq y_o$, $Fx_o \geq 0 \geq Fy_o$,	(H_2)
(34)	$Fy - Fx \leq A(u,v)(y-x)$, $x_{-1} \leq x \leq y \leq u \wedge v$.	
(35)	$x_o \geq y_o$, $Fx_o \leq 0 \leq Fy_o$,	(H_3)
(36)	$Fy - Fx \geq A(u,v)(y-x)$, $x_{-1} \leq x \leq y \leq u \wedge v$.	
(37)	$x_o \leq y_o$, $Fx_o \leq 0 \leq Fy_o$,	(H_4)
(38)	$Fy - Fx \leq A(u,v)(y-x)$, $u \vee v \leq x \leq y \leq y_o$.	

$x_n \leq x_{n+1} \leq y_{n+1} \leq y_n$, $Fx_n \geq 0 \geq Fy_n$, $n=0,1,\ldots$ | (C_1)

$x_n \geq x_{n+1} \geq y_{n+1} \geq y_n$, $Fx_n \geq 0 \geq Fy_n$, $n=0,1,\ldots$ | (C_2)

$x_n \geq x_{n+1} \geq y_{n+1} \geq y_n$, $Fx_n \leq 0 \leq Fy_n$, $n=0,1,\ldots$ | (C_3)

$x_n \leq x_{n+1} \leq y_{n+1} \leq y_n$, $Fx_n \leq 0 \leq Fy_n$, $n=0,1,\ldots$ | (C_4)

4.1. Theorem. Consider a nonlinear operator $F:D \subset X \to Y$, where X is a regular POTL-space and Y a POTL-space, three comparable points x_o, x_{-1}, y_o of D satisfying condition (29), and a mapping $A:D_1 \to B(X,Y)$, where D_1 is the set defined by (30). Let $(p_n)_{n\geq 0}$, $(q_n)_{n\geq 0}$ be two sequences of integers satisfying condition (2). Assume that hypothesis (H_i) is satisfied for an $i \in \{1,2,3,4\}$. Assume moreover that the linear operator $(-1)^i A(u,v)$ has an injective nonnegative continuous left subinverse for all $(u,v) \in D_1$. Then:

1°. The iterative algorithm

(39) $\qquad Fx_n + A_n(x_{n+1} - x_n) = 0$

(40) $\qquad Fy_n + A_n(y_{n+1} - y_n) = 0 \qquad n=0,1,\ldots$

(41) $\qquad A_n \in \{A(x_{p_n}, x_{q_n}), A(x_{q_n}, x_{p_n})\}$

is well defined (i.e. for any $n \in Z_+$ there are x_{n+1} and y_{n+1} satisfying (39) and (40)).

2°. Conclusion (C_i) holds.

3°. There exist two comparable points x^*, y^* in $\langle x_o, y_o \rangle$ such that $x^* = \lim\limits_{n\to\infty} x_n$, $y^* = \lim\limits_{n\to\infty} y_n$.

4°. If the operators $(-1)^i A_n$, $n=0,1,2,\ldots$ are inverse nonnegative then any solution of the equation $Fx=0$ in $\langle x_o, y_o \rangle$ belongs to $\langle x^*, y^* \rangle$ (i.e. $u \in \langle x_o, y_o \rangle$ and $Fu=0$ imply $u \in \langle x^*, y^* \rangle$).

Proof. We shall make the proof for the case i=1. Let B_o be a continuous nonsingular nonnegative left subinverse of $(-A_o)$ and let us consider the operator

$$H:[0,y_o-x_o]\to X\ ,\quad Hx=x-B_o(Fy_o-A_ox)\ .$$

It is easy to see that H is isotone and continuous. We also have:

$H(0)=-B_oFy_o\geq 0$,

$H(y_o-x_o)=y_o-x_o-B_oFx_o-B_o(Fy_o-Fx_o-A_o(y_o-x_o))\leq y_o-x_o-B_oFx_o\leq y_o-x_o$.

According to Kantorovich's theorem [8] the operator H has a fixed point $Hw=w\varepsilon[0,y_o-x_o]$. Taking $y_1=y_o-w$ we have

$$Fy_o+A_o(y_1-y_o)=0,\quad x_o\leq y_1\leq y_o\ .$$

Using (32) we deduce that

$$Fy_1=Fy_1-Fy_o-A_o(y_1-y_o)\leq 0\ .$$

Now let us define the operator

$$G:[0,y_1-x_o]\to X,\ Gx=x+B_o(Fx_o+A_ox)\ .$$

G is clearly continuous isotone and we have:

$G(0)=B_oFx_o\geq 0$,

$G(y_1-x_o)=y_1-x_o+B_oFy_1-B_o(Fy_1-Fx_o-A_o(y_1-x_o))\leq y_1-x_o+B_oFy_1\leq y_1-x_o$.

Applying again Kantorovich's theorem [8] we deduce the existence of a point $z\varepsilon[0,y_1-x_o]$ such that z=Gz. Taking $x_1=x_o+z$ it follows that

$$Fx_o+A_o(x_1-x_o)=0,\quad x_o\leq x_1\leq y_1\ .$$

Using the above relations and condition (32) we obtain

$$Fx_1=Fx_1-Fx_o-A_o(x_1-x_o)\geq 0\ .$$

Proceeding by induction we can show that there exist two sequences $(x_n)_{n\geq 1}$ and $(y_n)_{n\geq 1}$ satisfying (C_1). The space X being regular it follows that there exist x^*, $y^*\varepsilon X$ such that $x^*=\lim_{n\to\infty} x_n^*$, $y^*=\lim_{n\to\infty} y_n$. We have obviously $x^*\leq y^*$.

If $x_o\leq u\leq y_o$ and Fu=0 then we can write

$$A_o(y_1-u)=A_oy_o-Fy_o-A_ou=A_o(y_o-u)-(Fy_o-Fu)\leq 0$$

and

$$A_o(x_1-u)=A_ox_o-Fx_o-A_ou=A_o(x_o-u)-(Fx_o-Fu)\geq 0.$$

If the operator $(-A_o)$ is inverse nonnegative then it follows that $x_1 \leq u \leq y_1$. Proceeding by induction we deduce that $x_n \leq u \leq y_n$ holds for all n. Hence $x^* \leq u \leq y^*$ ∎

To complete the statement of the above theorem we shall give some natural conditions under which the points x^* and y^* are solutions of the equation $Fx=0$.

4.2. Proposition. Under the hypothesis of Theorem 4.1 suppose that F is continuous at x^* and y^*. If one of the following conditions is satisfied:

(i) X is normal and there exists an operator $T \in L(X,Y)$, having a continuous nonnegative inverse, such that $A_n \leq T$ for sufficiently large n;

(ii) Y is normal and there exists an operator $S \in L(X,Y)$ such that $A_n \leq S$ for sufficiently large n;

(iii) the operators A_n, $n=0,1,2,\ldots$ are equicontinuous.

Then $Fx^*=Fy^*=0$ ∎

The proof of this proposition is very simple and will be omitted (see [17]).

Let us note that if F has a linear continuous Gâteaux derivative $F'(x)$ at each point $x \in \langle x_{-1}, y_o \rangle$ and if the mapping $F': \langle x_{-1}, y_o \rangle \to B(X,Y)$ is isotone, then conditions (32) and (34) are satisfied taking respectively

$$A(u,v)=F'(u \vee v) \quad , \quad (u,v) \in D_1 \tag{42}$$

$$A(u,v)=F'(u \wedge v) \quad , \quad (u,v) \in D_1 \; . \tag{43}$$

If F' is antitone then (36) is fulfilled for A given by (43), and (38) is fulfilled for A given by (42). With the choice (3), (42) the results of Theorem 4.1 and Proposition 4.2 constitute a slight improvement of the result of M.H.Wolfe [25].

We also note that conditions (32) and (34) are satisfied if A is a divided difference of F on $\langle x_{-1}, y_o \rangle$ (i.e. $A(u,v)(u-v)=Fu-Fv$) which is isotone in each argument, while conditions (36) and (38) are satisfied if A is a divided difference antitone in each argument. This shows that the results contained in Theorem 4.1 and Proposition 4.2 represent a generalization of the result obtained by J.W.Schmidt and H. Leonhardt [19] concerning the secant method (see also [17] and [21]).

Now let us make some remarks on the regularity assumption of the space X appearing in the statement of Theorem 4.1. This assumption was essentially used in proving that the iterative procedure (39)-(41) is well defined and that the sequences produced by it are convergent. In Proposition 4.2 we have given some sufficient conditions under which the limits of these sequences are roots of the equation $Fx=0$. We have already mentioned in Section 2 that the regularity condition is rather restrictive. In some cases the existence of the solution can be proved by other means without this assumption and iterative procedures are applied for enclosing the solution (see [1]). In the following theorem we shall show that an "explicit version" of the iterative procedure (39)-(41) can be used to this effect.

4.3. Theorem. Consider a nonlinear operator $F:D\subset X\to Y$, where X and Y are POL-spaces and let x_{-1},x_o,y_o be three comparable points of D satisfying condition (29). Consider also a mapping $A:D_1\to L(X,Y)$ where D_1 is the set defined by (30). Let $(p_n)_{n\geq 0}$ and $(q_n)_{n\geq 0}$ be two sequences of integers satisfying condition (2) and let i be a fixed integer between 1 and 4. Assume the operator $(-1)^i A(u,v)$ has a nonnegative subinverse for any $(u,v)\in D_1$ and hypothesis (H_i) is satisfied.

Then the iterative algorithm:

(44) $$x_{n+1}=x_n-B_nFx_n$$

$$n=0,1,2,\ldots$$

(45) $$y_{n+1}=y_n-B'_nFy_n$$

where $(-1)^iB_n$, $(-1)^iB'_n$ are nonnegative subinverses of $(-1)^iA_n$, $(-1)^iA'_n$ $\in\{(-1)^iA(x_{p_n},x_{q_n}),(-1)^iA(x_{q_n},x_{p_n})\}$, generates two sequences $(x_n)_{n\geq 0}$, $(y_n)_{n\geq 0}$ satisfying conclusion (C_i). Moreover for any solution $u\in$ $\in\langle x_o,y_o\rangle$ of the equation $Fx=0$ we have

(46) $$u\in\langle x_n,y_n\rangle\ , \qquad n=0,1,2,\ldots$$

Proof. We shall prove the theorem for i=1. In this case we have:

(47) $$B_o\leq 0,\ B'_o\leq 0,\ I\geq A_oB_o\ ,\ I\geq B_oA_o\ ,\ I\geq A'_oB'_o\ ,\ I\geq B'_oA'_o\ .$$

From (31), (32), (44), (45) and (47) it follows that:

$$y_o-y_1=B'_oFy_o\geq 0\ ,$$

$$y_1-x_o=y_o-x_o-B'_oFy_o\geq y_o-x_o-B'_o(fy_o-Fx_o)\geq B'_o(A'_o(y_o-x_o)-(Fy_o-Fx_o))\geq 0,$$

$$x_1-x_o=-B_oFx_o\geq 0\ ,$$

$$y_o-x_1=y_o-x_o+B_oFx_o\geq y_o-x_o-B_o(Fy_o-Fx_o)\geq B_o(A_o(y_o-x_o)-(Fy_o-Fx_o))\geq 0.$$

Hence $x_1,y_1\in[x_o,y_o]$. Using again (31), (32), (44), (45) and (47) we obtain:

$$Fy_1=Fy_1+A'_o(y_o-y_1-B'_oFy_o)=Fy_1-A'_oB'_oFy_o+A'_o(y_o-y_1)\leq Fy_1-Fy_o+A'_o(y_o-y_1)\leq 0\ ,$$

$$Fx_1=Fx_1-A_o(x_1-x_o+B_oFx_o)=Fx_1-A_oB_oFx_o-A_o(x_1-x_o)\geq Fx_1-Fx_o-A_o(x_1-x_o)\geq 0,$$

$y_1-x_1 \geq y_1-x_1+B_0'Fx_1=y_0-x_1-B_0'(Fy_0-Fx_1) \geq B_0'(A_0'(y_0-x_1)-(Fy_0-Fx_1)) \geq 0$.

Thus we have proved that $x_0 \leq x_1 \leq y_1 \leq y_0$ and $Fx_1 \leq 0 \leq Fy_1$. Proceeding by induction we deduce that (C_1) is satisfied.

Finally, if $u \in [x_0,y_0]$ and $Fu=0$ then we may write:

$$y_1-u=y_0-u-B_0'Fy_0+B_0'Fu \; B_0'(A_0'(y_0-u)-(Fy_0-Fu)) \geq 0 \; ,$$

$$u-x_1=u-x_0+B_0Fx_0-B_0Fu \; B_0(A_0(u-x_0)-(Fu-Fx_0)) \geq 0 \; .$$

Hence $x_1 \leq u \leq y_1$ and, by induction, $x_n \leq u \leq y_n$ for $n=1,2,\ldots$ ∎

REFERENCES

1. ALEFELD, G.: Monotone Regula-Falsi-ähnliche Verfahren bei nichtkonvexen Operatorgleichungen, Beiträge zur Numerische Mathematik, 8(1979).
2. BOSARGE, W.E. and FALB, P.L.: A multipoint method of third order, J.Optimiz.Theory Appl., 4(1969), 156-166.
3. BOSARGE, W.E; and FALB, P.L.: Infinit dimensional multipoint methods and the solution of two point boundary value problems, Numer.Math., 14(1970), 264-286.
4. DENNIS, J.E.: On the Kantorovich hypothesis for Newton's method, SIAM J.Numer.Anal., 6, 3(1969), 493-507.
5. DENNIS, J.E.: Toward a unified convergence theory for Newton--like methods, in Nonlinear Functional Analysis and Applications, L.B.Rall, Ed., Academic Press, New York, 1971.
6. DEUFLHARD, P. and HEINDL, G.: Affine invariant convergence theorems for Newton's method and extensions to related methods, SIAM J.Numer.Anal., 16(1979), 1-10.
7. GRAGG, W.B. and TAPIA, R.A.: Optimal error bounds for the Newton--Kantorowich theorem, SIAM J.Numer.Anal., 11(1974), 10-13.
8. KANTOROVICH, L.V.: The method of succesive approximations for functional equations, Acta Math., 71(1939), 63-97.
9. LAASONEN, P.: Ein überquadratisch konvergenter iterativer Algorithmus. Ann.Acad.Sci.Fenn.Ser.A. I, 450(1969), 1-10.
10. MIEL, G.J.: An updated version of the Kantorovich theorem for Newton's method. Technical Summary Report, Mathematical Research Center, University of Wisconsin Madison, 1980.
11. ORTEGA, J.M. and RHEINBOLDT, W.C.: Monotone iteration for nonlinear equations with application to Gauss-Sudel methods, SIAM J.Numer.Anal., 4(1967), 171-190.
12. POTRA, F.-A.: On the aposteriori error estimates for Newton's

method, Preprint Series in Mathematics, INCREST, Bucharest, no. 19/1981.

13. POTRA, F.-A.: A general iterative procedure for solving nonlinear equations in Banach spaces, Preprint Series in Mathematics, INCREST, Bucharest, no.37/1981.
14. POTRA, F.-A. and PTÁK, V.: Sharp error bounds for Newton's method, Numer.Math., 34(1980), 63-72.
15. POTRA, F.-A. and PTÁK, V.: On a class of modified Newton processes, Numer.Func-Anal. and Optimiz., 2, 1(1980), 107-120.
16. POTRA, F.-A. and PTÁK, V.: A generalization of Regula Falsi Numer.Math., 36(1981), 333-346.
17. POTRA, F.-A. and SCHMIDT, J.W.: On a class of iterative procedures with monotonous convergence, Preprint Series in Mathematics, INCREST, Bucharest, no.3/1982.
18. SCHMIDT, J.W.: Eine Übertragung der Regula Falsi auf Gleichungen in Banachraum, I, II, Z.Angew.Math.Mech., 43(1963), 1-8, 97-110.
19. SCHMIDT, J.W. and LEONHARDT, H.: Eingrenzung von Lösungen mit Hilfe der Regula Falsi, Computing, 6(1970), 318-329.
20. SCHMIDT, J.W. and SCHWETLICK, H.: Ableitungsfreie Verfahren mit höherer Konvergenzgeschwindigkeit, Computing, 3(1968), 215--226.
21. SCHNEIDER, N.: Monotone Einschliessung durch Verfahren von Regula-falsi-Typ unter Verwendung eines verallgemeinerten Steigungsberiffes, Computing, 26(1981), 33-44.
22. TRAUB, J.F.: Iterative methods for the solution of equations, Prentice Hall, Englewood Cliffs, New Jersey, 1964.
23. VANDERGRAFT, J.S.: Newton's method for convex operators in partially ordered spaces, SIAM J.Numer.Anal., 4(1967), 406-432.
24. WOLFE, M.A.: Extended iterative methods for the solution of operator equations, Numer.Math., 31(1978), 153-174.
25. WOLFE, M.A.: On the convergence of some methods for determining zeros of order-convex operators, Computing, 26(1981), 45-56.

ADI-METHODS FOR NONLINEAR VARIATIONAL INEQUALITIES OF EVOLUTION

Ulrich Hornung

1. The Initial Boundary Value Problem

A large class of nonlinear diffusion problems can be written in the form

$$(1.1)\quad \begin{cases} \partial_t G(t,x,v) = \nabla(K(t,x,v)\nabla v + Y(t,x,v)) - F(t,x,v) & \text{on } Q \\ v = D(t,x) & \text{on } \Sigma_D, \\ q_\nu = E(t,x,v) & \text{on } \Sigma_N, \\ v \geq 0,\ q_\nu \leq 0,\ v\cdot q_\nu = 0 & \text{on } \Sigma_U, \\ G(0,x,v) = u_0(x) & \text{on } \Omega. \end{cases}$$

Here we assume that Ω is a bounded domain in $\mathbb{R}^n$ with a smooth boundary $\Gamma = \Gamma_D + \Gamma_N + \Gamma_U$, and we define $Q = [0,T] \times \Omega$, $\Sigma_i = [0,T] \times \Gamma_i$ for $i = D,N$, and U. The functions $D\colon \Sigma_D \to \mathbb{R}$, $E\colon \Sigma_N \times \mathbb{R} \to \mathbb{R}$, $F\colon Q \times \mathbb{R} \to \mathbb{R}$, $G\colon Q \times \mathbb{R} \to \mathbb{R}$, $Y\colon Q \times \mathbb{R} \to \mathbb{R}^n$, $K\colon Q \times \mathbb{R} \to \mathbb{R}^{n,n}$, and $u_o\colon \Omega \to \mathbb{R}$ are the data of the problem. The outer normal vector on Γ is ν and $q_\nu = -(K\nabla v + Y)\cdot\nu$ is the normal flow.

Problems of this type occur in heat conduction, transport of solutes, flow of fluids or gases in porous media, melting problems, chemical engineering, etc. A survey together with existence and uniqueness theorems for water flow in porous media can be found in Hornung (2). Theoretical results are also contained in Alt/Luckhaus (1). A weak but general existence theory is developped in Hornung (4).

The main qualitative assumptions on the data of the problem are that G is monotone with respect to v, that K and Y are Lipschitz functions of v and uniformly bounded, and K is uniformly positive definite. The differential equation in (1.1) is parabolic if G is strongly monotone with respect to v, but in general it is of parabolic-elliptic type. Since the equation degenerates whenever G is constant, it is essential to apply fully implicit methods with respect to time when solving the problem numerically. If the exact solution is smooth, backwards differentiation formulas of higher order may be used, but otherwise the simple backwards Euler method is appropriate. Though it is of order 1 this method has the advantage to be conservative, i.e. it has an exact mass or energy balance, resp. (cf. Hornung/ Messing (5)). It is known that the transversal line method, i.e. the fully implicit discretization of the time variable, is $L^1(\Omega)$-convergent uniformly on every finite interval $[0,T]$ of order 1/2 (cf. Hornung (4)), in special cases it is $L^2(Q)$-convergent (cf. Hornung (3)). For Dirichlet and Neuman boundary data it is shown in Messing (7) that the $L^2(Q)$-order of convergence is 1/2 and that it is 1 if the function v has a distributional derivative $\partial_t v$ in $L^2(Q)$. All these statements hold for the discretization of the function $u = G(t,x,v)$.

2. The Nonlinear System

If an approximation v^o of $v(t^o)$ has been computed, the implicit Euler method consists in solving the nonlinear elliptic variational inequality

$$(2.1)\quad \begin{cases} \frac{1}{\Delta t}(G(t,x,v) - G(t^o,x,v^o)) = \nabla(K(t,x,v)\nabla v + Y(t,x,v)) - F(t,x,v) & \text{in } \Omega, \\ v = D(t,x) & \text{on } \Gamma_D, \\ q_\nu = E(t,x,v) & \text{on } \Gamma_N, \\ v \geq 0,\ q_\nu \leq 0,\ v\cdot q_\nu = 0 & \text{on } \Gamma_U, \end{cases}$$

for v, which gives an approximation of $v(t)$, $t = t^o + \Delta t$. The weak formulation of this problem is

$$(2.2)\quad \langle A(v,v) + N(v), w - v\rangle \geq 0 \quad \text{for all } w \in M,$$

where we define

$$\langle A(u,v),\Phi\rangle = \int_\Omega (K(t,x,u)\nabla v + Y(t,x,u))\cdot\nabla\Phi\, dx$$

$$\langle N(v),\Phi\rangle = \int_{\Omega} \left[\frac{1}{\Delta t} (G(t,x,v) - G(t^o,x,v^o)) + F(t,x,v)\right]\cdot\Phi \, dx + \int_{\Gamma_N} E(t,x,v)\cdot\Phi \, d\Gamma$$

and

$$M = \{v \in H^1(\Omega) : v = D(t,x) \text{ on } \Gamma_D \text{ and } v \geq 0 \text{ on } \Gamma_U\}.$$

Existence of a weak solution follows from the fact that $v \to A(v,v) + N(v)$ is a pseudo-monotone coercive operator from $H^1(\Omega)$ into its dual space (cf. Hornung (4)).

In the sequel for simplicity we neglect Y and assume that K is real valued. Then in two space dimensions the operator A has the natural splitting

$$A(u,v) = H(u,v) + V(u,v),$$

where we have

$$\langle H(u,v),\Phi\rangle = \int_{\Omega} K(u)\partial_x v\cdot\partial_x\Phi \, dx\, dy$$

and

$$\langle V(u,v),\Phi\rangle = \int_{\Omega} K(u)\partial_y v\cdot\partial_y\Phi \, dx\, dy.$$

A finite nonlinear system is obtained from (2.1) or (2.2) by discretization of the space variable. This can be done either using finite elements or finite differences. The main advantage of the finite element method is that irregular geometries can be treated in an elegant way, whereas finite differences are very easy to implement for problems with a simple geometrical structure. In any case the continuous problem (2.2) is replaced by the discrete problem

(2.3) $\langle A_n(v_n,v_n) + N_n(v_n), w_n - v_n\rangle \geq 0$ for all $w_n \in M_n$,

where M_n is a closed convex set in a finite dimensional space. Here again in two dimensions there is a natural splitting

$$A_n(u_n,v_n) = H_n(u_n,v_n) + V_n(u_n,v_n).$$

For instance on a rectangular grid of equidistant mesh sizes Δx and Δy the expression $\langle H_n(u_n,v_n),\Phi_n\rangle$ is built up by terms of the form

$$\frac{1}{(\Delta x)^2}\,(K_{i+1/2,j}(u)\cdot(v_{i+1,j}-v_{i,j}) - K_{i-1/2,j}(u)\cdot(v_{i,j}-v_{i-1,j}))$$

and in $\langle V_n(u_n,v_n),\Phi_n\rangle$ we use

$$\frac{1}{(\Delta y)^2}\,(K_{i,j+1/2}(u)\cdot(v_{i,j+1}-v_{i,j}) - K_{i,j-1/2}(u)\cdot(v_{i,j}-v_{i,j-1})).$$

In the following we drop the index n, since (2.2) and (2.3) are of the same formal structure.

For a linear system of equations

(2.4) $\quad Hv + Vv + N = 0$

the Peaceman-Rachford ADI-method uses the iteration process

$$(\alpha I + H)v^{k+1/2} = (\alpha I - V)v^k - N$$

$$(\beta I + V)v^{k+1} = (\beta I - H)v^{k+1/2} - N$$

with acceleration parameters $\alpha,\beta > 0$. If H and V are linear, but N is a nonlinear diagonal operator, this method can be modified in order to solve

$$Bv = Hv + Vv + N(v) = 0$$

by using the method

$$(\alpha I + H + N'(v^k))v^{k+1/2} = (\alpha I - V + N'(v^k))v^k - N(v^k)$$

$$(\beta I + V + N'(v^{k+1/2}))v^{k+1} = (\beta I - H + N'(v^{k+1/2}))v^{k+1/2} - N(v^{k+1/2})$$

where $N'(v)$ denotes the differential of N. Since this scheme is equivalent to

$$v^{k+1/2} = v^k - (\alpha I + H + N'(v^k))^{-1}B(v^k)$$

$$v^{k+1} = v^{k+1/2} - (\beta I + V + N'(v^{k+1/2}))^{-1}B(v^{k+1/2})$$

this is obviously a variant of the Peaceman-Rachford-Newton method (cf. Ortega/Rheinboldt (8)).

Generalizing the underlying idea we solve the problem

$$(2.5) \qquad \langle H(v,v) + V(v,v) + N(v),\ w - v\rangle \geq 0 \quad \text{for all } w \in M$$

by the ADI-iteration method

$$(2.6) \quad \begin{cases} \langle \alpha I + H(v^k,.) + N'(v^k), w - v^{k+1/2}\rangle \geq \\ \qquad \langle \alpha I - V(v^k,.) + N'(v^k) - N(v^k), w - v^{k+1/2}\rangle, \\ \langle \beta I + V(v^{k+1/2},.) + N'(v^{k+1/2}), w - v^{k+1}\rangle \geq \\ \qquad \langle \beta I - H(v^{k+1/2},.) + N'(v^{k+1/2}) - N(v^{k+1/2}), w - v^{k+1}\rangle \end{cases}$$

$$\text{for all } w \in M.$$

Each of the two sets of inequalities can be treated very quickly, since they amount in solving one-dimensional linear elliptic problems, where the unilateral boundary conditions can be checked and satisfied easily by trial and error.

3. Iteration Parameters

The efficiency of the iteration method depends strongly on the initial guess and on the choice of the acceleration parameters. Since we are dealing with an evolution problem, it is natural to use v^o as an initial guess for the approximation of $v(t^o + \Delta t)$. Of course this can be improved by extrapolation from some recent time steps. The extrapolation is the better the smaller the time step size Δt is chosen. Therefore convergence problems are avoided by a step size control procedure. The general philosophy of this procedure is to keep the step size Δt small enough in order to prevent the iteration from divergence while at the same time yielding some prescribed order of variation of v within one time step. Thus some required accuracy is assured with a reasonable amount of effort. This strategy is described in more detail in Hornung/Messing (5).

Since the theory of optimal acceleration parameters for nonlinear ADI-methods is very difficult, linear problems are used as a guideline. If the function K does not depend on the time and space variables, we may assume without loss of generality that it is a constant, since Kirchhoff's transformation may be applied. In this case (2.2) and (2.3) can be written as

$$(3.1) \quad \langle H + V + N(v), w - v \rangle \geq 0 \quad \text{for all } w \in M,$$

where H and V are linear operators, and N is nonlinear diagonal. Here (2.4) can be considered as a model problem, for which optimal acceleration parameters are obtained according to well known analysis (cf. Wachspress (9)). These parameters depend on the type of the boundary conditions and on some prescribed length of iteration cycles. Since this theory is applicable for Dirichlet and/or Neuman boundary conditions, some choice has to be made on the part Γ_U of Γ. Only a rough idea on this is sufficient, since the acceleration parameters are not very sensitive to a change of the type on a small part of the boundary. In practice the number of iteration steps is small, therefore short cycles may be chosen, e.g. those of length 2 or 4.

The strategy described above worked very well for various test problems. Up to now no reasonable way was found to determine good parameters for problems with a function K depending on the space variable. This is an important problem, since for water flow in nonhomogeneous porous media the conductivity K may vary over several orders of magnitude.

4. Examples

Several test cases have been treated in the way described in the preceeding paragraphs. A simple but typical example is given in the following situation. The differential equation

$$(4.1) \quad \partial_t G(v) = \Delta v$$

with the monotone function

$$(4.2) \quad G(v) = 1/2\, v_+^2$$

is solved in a rectangle $\Omega \subset \mathbb{R}^2$. A radial symmetric solution can easily be found using polar coordinates. For $v > 0$ the differential equation reads in this case

(4.3) $v \cdot \partial_t v = \partial_{rr} v + 1/r \partial_r v.$

Therefore Boltzmann's transform $v = z(s)$, $s = r/\sqrt{t}$ leads to the ordinary differential equation

(4.4) $z'' = -(1/s + 1/2 \cdot s \cdot z) \cdot z'$,

where the prime denotes differentiation with respect to s. If z is a solution of (4.4) with initial conditions

(4.5) $z(1) = 0, \quad z'(1) = \lambda,$

a solution of (4.1), (4.2) is given by

$$v(t,x,y) = \begin{cases} z(+\infty) & \text{if } t \leq 0, \\ \lambda \log s \quad \text{for } 0 < s \leq 1 \\ z(s) \quad \text{for } s > 1 \end{cases} \quad \text{if } t > 0.$$

This solution represents saturated/unsaturated water flow in a homogeneous porous medium from a point source at $r = 0$ with constant discharge. Figures 1 and 2 show the functions v and $G(v)$, resp., at two

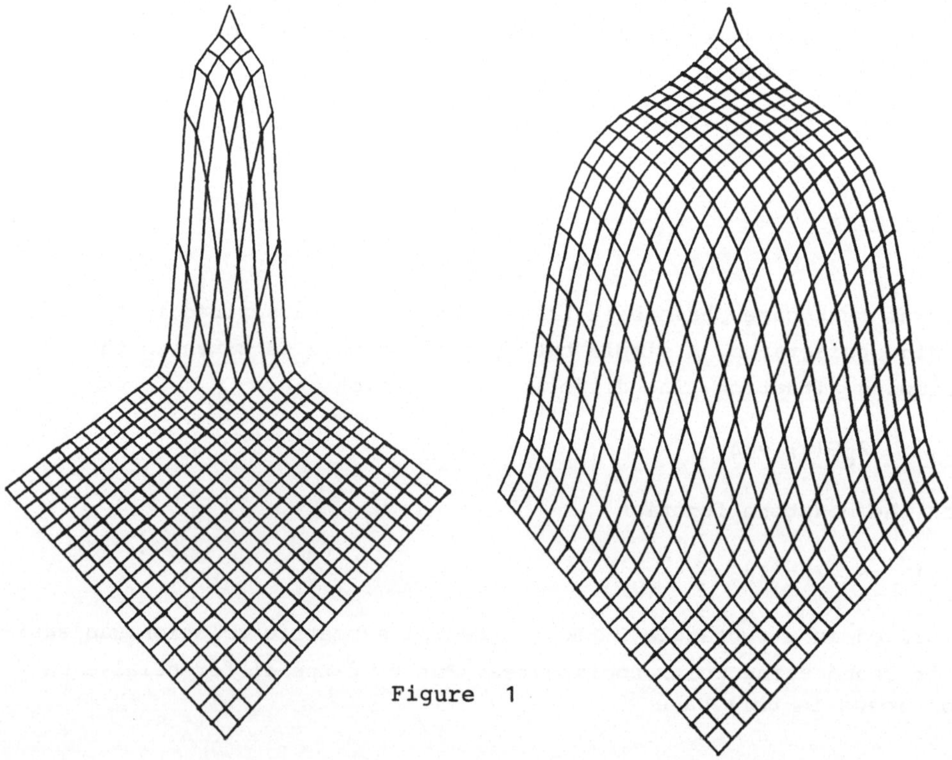

Figure 1

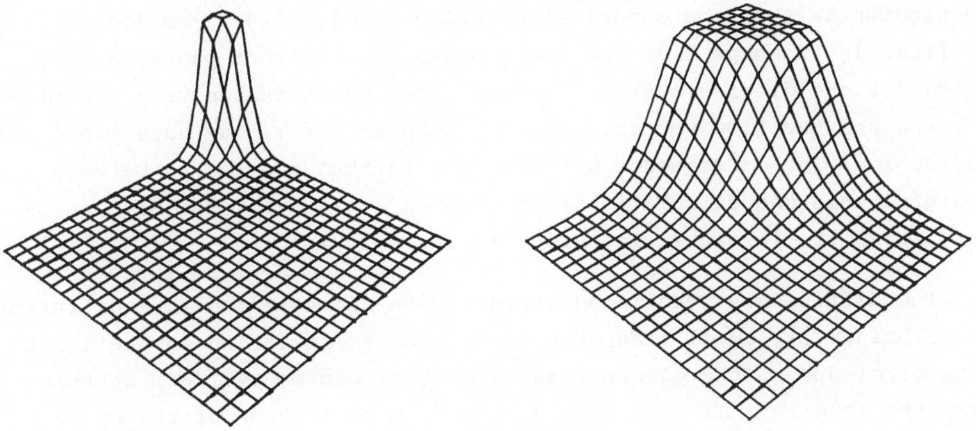

Figure 2

different instants of time. Here the source is close to one of the corners outside of Ω.

The main difficulty in the numerical solution of this problem is that at the beginning there is a very sharp circular wetting front which propagates with speed $\sqrt{t}$. As an illustration in Figure 3 equipotentials, i.e. lines of constant v, are plotted for $t = 0.01$. The

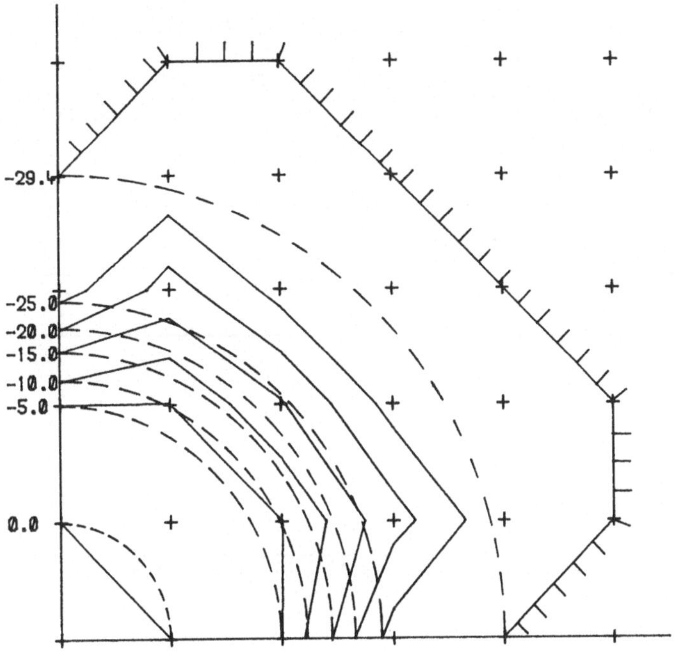

Figure 3

equipotentials of the numerical solution - solid lines - differ significantly in shape from the exact ones. If the front passes a node used for the discretization, the numerical solution has very low precision and the flow vector, i.e. the gradient of v, may have wrong direction. The consequence of this fact is that one has to be very careful when the calculations are used as a basis for the simulation of transport of solutes in porous media.

The difficulties mentioned above can be demonstrated in a different way. Using very small time step sizes and extrapolation to the limit the error due to the discretization of time can be avoided. In this way the longitudinal line method, i.e. discretization of the space variable only, is realized. Now the discrete $L^{\infty}(\Omega)$ - error is a function of time, which is plotted in Figure 4 for different spatial mesh sizes Δx. This figure shows that a reduction of Δx essentially changes only the time scale of the error curve rather than decreasing the maximum error. This means that in the case of sharp fronts it is practically impossible to reduce the L^{∞}-error to a small tolerance with reasonable effort (cf. Hornung/Messing (6)), while the L^2-error is of order 1.

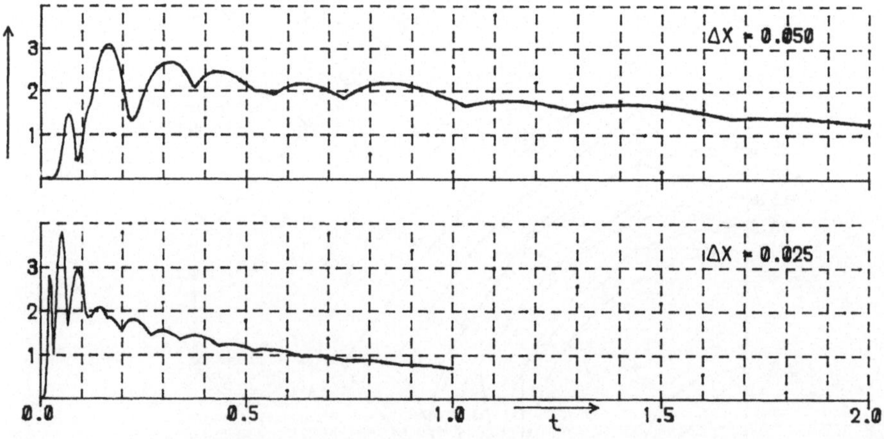

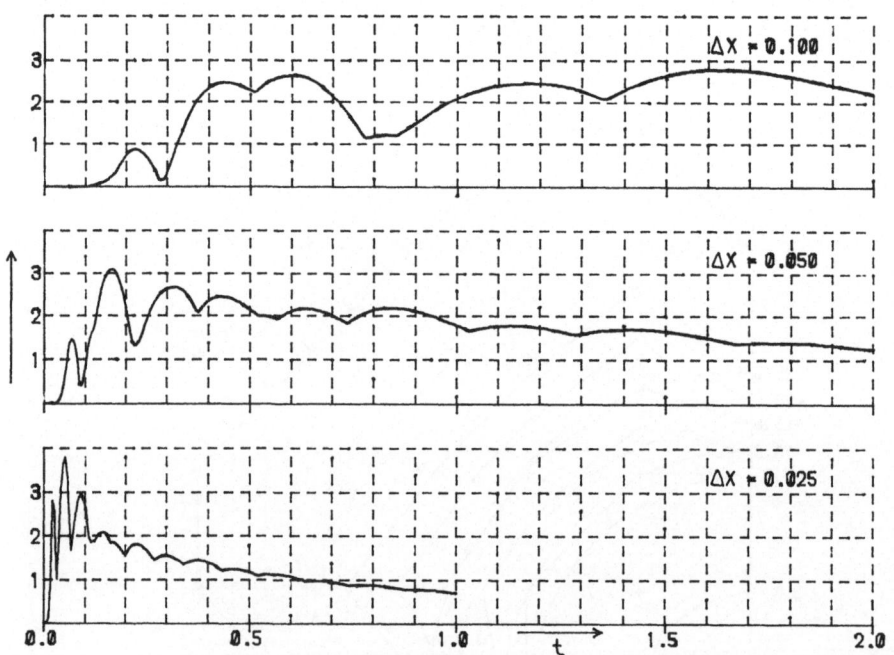

Figure 4

5. Conclusion

ADI-iteration is an efficient a means for solving nonlinear systems that occur as conservative discretization schemes for variational inequalities of evolution.

6. References

1 Alt, H.W., Luckhaus, S.: Quasi-linear elliptic-parabolic differential equations. Sonderforschungsbereich 123, Preprint 136, Heidelberg, 1982.

2 Hornung, U.: A unilateral boundary value problem for unsteady waterflow in porous media. Gorenflo, R., Hoffmann, K.H. (Eds.): "Variational Methods and Ill-Posed Problems". Bibliographisches Institut (1982).

3 Hornung, U.: Convergence of the transversal line method for a parabolic-elliptic equation. To appear.

4 Hornung, U.: A weak existence theorem for nonlinear diffusion problems. To appear.

5 Hornung, U., Messing, W.: Simulation of two-dimensional saturated/unsaturated flows with an exact water balance. Verruijt, A., Barends, F.B.J. (Eds.):"Flow and Transport in Porous Media". Proceedings of EUROMECH 143/Delft. Balkema, Rotterdam (1981), 91-96.

6 Hornung, U., Messing, W.: Truncation errors in the numerical solution of horizontal diffusion in saturated/unsaturated media. Advances in Water Resources (1982).

7 Messing, W.: Konvergenz-Ordnung von Diskretisierungsverfahren für nichtlineare parabolisch-elliptische Differentialgleichungen. Dissertation, Münster, 1982.

8 Ortega, J.M., Rheinboldt, W.C.: Iterative Solution of Nonlinear Equations in Several Variables. Academic Press (1970).

9 Wachspress, E.: Iterative Solution of Elliptic Systems. Prentice Hall (1966).

Acknowledgement This work was supported by the Deutsche Forschungsgemeinschaft (Grant Ho 782/2).

RELAXATION METHODS FOR THE COMPUTATION OF THE SPECTRAL NORM

Gerhard Kolb
Rechenzentrum der Universität Mannheim
D-6800 Mannheim, Postfach

Wilhelm Niethammer
Institut für Praktische Mathematik
Universität Karlsruhe
D-7500 Karlsruhe, Englerstr. 2

Abstract:

Relaxation methods of H.R. Schwarz and A. Ruhe are used for the computation of the spectral norm σ_1 of an arbitrary $m \times n$-matrix A (i.e., σ_1 is the maximal singular value of A). σ_1 is eigenvalue and spectral radius of a symmetric weakly two-cyclic matrix $\hat{A}$. Using this fact, results on the optimal (asymptotic) relaxation factor ω_o are derived. Further it is shown that using ω_o from the beginning of the algorithm often prevents a rapid convergence. A strategy for the choice of ω is presented.

1. Introduction:

Let A be a real $m \times n$-matrix with $m \geq n$ and let

(1.1) $$\sigma_1^2 \geq \sigma_2^2 \geq \dots \geq \sigma_n^2 \geq 0$$

be the eigenvalues of A^TA; then σ_1 is the spectral norm $||A||_2$ of A, whereas $\sigma_1 \geq \dots \geq \sigma_n \geq 0$ are the singular values of A. $\sigma_1, \dots, \sigma_n$ together with the corresponding eigenvectors of A^TA, resp. AA^T, determine the singular value decomposition (svd) of A (cf. [1]). There are algorithms for computing the svd ([2], [3]). But there may be applications where it is sufficient to know σ_1 and the corresponding eigenvectors $x \in \mathbb{R}^m$ of AA^T and $y \in \mathbb{R}^n$ of A^TA (x and y are called left and right singular vectors of A, corresponding to σ_1). We propose an algorithm for computing σ_1, x and y; more details are given in [4].

At first we observe that σ_1 is the spectral radius of the $(m+n)\times(m+n)$ matrix

$$\hat{A} := \begin{bmatrix} 0 & A \\ A^T & 0 \end{bmatrix} \tag{1.2}$$

($\hat{A}$ is used too in [2]). Further, σ_1 is eigenvalue of $\hat{A}$. To σ_1 corresponds the eigenvector $\hat{z} := (x,y)^T$, where x and y are left resp. right singular vectors of A, i.e., $\lambda = \sigma_1$ and $z = \hat{z}$ are solutions of the eigenvalue equation

$$C(\lambda)z := (\lambda I - \hat{A})z = 0. \tag{1.3}$$

Now relaxation methods, developed by H.R. Schwarz ([6], [7]) and A. Ruhe ([5]) for the computation of the smallest eigenvalues of the general eigenvalue problem, are applied for the computation of σ_1 and $\hat{z}$.

In Section 2, the algorithm is formulated; a sequence $\{\lambda_k\}_{k\geq o} \subset \mathbb{R}$ and a sequence $\{z_k\}_{k\geq o} \subset \mathbb{R}^{m+n}$ are generated. In Section 3 the convergence of $\{\lambda_k\}$ to an eigenvalue of $\hat{A}$ and of $\{z_k\}$ to a corresponding eigenvector is proven. For the consideration of the asymptotic convergence behavior of $\{z_k\}$ in Section 4, it is important, that $\hat{A}$ is weakly two-cyclic ([8]), p. 39). It follows that convergence can only happen to σ_1, resp. $\hat{z}$. If

$$\mu := \max\{\sigma_i/\sigma_1 \mid \sigma_i < \sigma_1\}, \tag{1.4}$$

then the optimal relaxation factor ω_{opt} is given in terms of μ by the well-known formula. A value of μ near 1 yields an ω_{opt} near 2. In Section 5, under some constraints, the inequality $\lambda_{k+1}/\lambda_k \leq 2/\omega$ is derived. I.e., if $\lambda_o \ll \sigma_1$ and ω_{opt} is near 2, then, if ω_{opt} is used from the beginning of the iteration, the monotonically increasing sequence $\{\lambda_k\}$ converges very slowly towards σ_1. In Section 6, a strategy for choosing ω is presented.

2. The Algorithm SOR

One step of the iteration goes as follows: Given z_k, we set λ_k to be the Raleigh quotient of z_k and $\hat{A}$. Now, the nonlinear equation (1.3) is replaced by the linear equation $(\lambda_k I - \hat{A})z = 0$. Applying one usual SOR step (Successive Over Relaxation) to this linear equation yields z_{k+1} from z_k. In more detail: Let $C(\lambda) := \lambda I - \hat{A} =: \lambda I - L - U$ with lower and upper triangular matrices L and U; from (1.2) we get

$$L = \begin{bmatrix} 0 & 0 \\ A^T & 0 \end{bmatrix}, \qquad U = \begin{bmatrix} 0 & A \\ 0 & 0 \end{bmatrix}. \tag{2.1}$$

If $\lambda_k \neq 0$, the usual SOR step is given by

$$(\lambda_k I - \omega L) z_{k+1} = ((1-\omega)\lambda_k I + \omega U) z_k \qquad \text{or}$$

$$z_{k+1} = T(\omega, \lambda_k) z_k \qquad \text{with} \tag{2.2}$$

$$T(\omega, \lambda_k) := (\lambda_k I - \omega L)^{-1} ((1-\omega)\lambda_k I + \omega U), \tag{2.3}$$

We get the following formulas for the

<u>Algorithm SOR</u>: Let $z_o \neq 0$ and $\omega \in (0,2)$ (the choice of z_o and ω will be discussed later)..
For k = 0,1,... do

$$z_k := z_k / \| z_k \|_2 \ ; \tag{2.4}$$

$$\lambda_k := z_k^T \hat{A} z_k \ ; \tag{2.5}$$

$$z_{k+1} = T(\omega, \lambda_k) z_k \ . \tag{2.6}$$

If some stop criterion isn't satisfied, go to (2.4) with k:= k + 1.
Using (1.2) and $z_k =: (x_k, y_k)^T$ with $x_k \in \mathbb{R}^m$, $y_k \in \mathbb{R}^n$, the formulas (2.4) to (2.6) can be simplified:

$$x_k := x_k / \| z_k \|_2 , \qquad y_k := y_k / \| z_k \|_2 \ ; \tag{2.4'}$$

$$\lambda_k := 2 x_k^T A y_k \ ; \tag{2.5'}$$

$$x_{k+1} = (1-\omega) x_k + \omega A y_k / \lambda_k \ ; \tag{2.6a'}$$

$$y_{k+1} = (1-\omega) y_k + \omega A^T x_{k+1} / \lambda_k \ . \tag{2.6b'}$$

We see from (2.6)' that for $\omega = 1$ SOR simplifies to vector iteration applied to $A^T A$ (without an explicit computation of $A^T A$).

3. Convergence

The following lemma follows the lines of Lemma 3.1 in [5]; here $\| \cdot \|$ means the Euclidean norm.

Lemma 1. Let $\| z_k \|_{k>o}$ and $\{\lambda_k\}_{k>o}$ be generated by SOR. Then with $p_k := z_k - z_{k+1}$ there holds

$$\lambda_{k+1} - \lambda_k = \frac{(2-\omega)}{\omega} \lambda_k \frac{\| p_k \|^2}{\| z_{k+1} \|^2} \tag{3.1}$$

<u>Proof</u>: Let $C_k := \lambda_k I - \hat{A}$. Then from (2.3) we get $T(\omega,\lambda_k) =: M_k^{-1} N_k$ with

$$M_k := \lambda_k I/\omega - L, \qquad N_k := (1-\omega)\lambda_k I/\omega + U$$

$$M_k + M_k^T = \lambda_k I - L - U + (2-\omega)\lambda_k I/\omega \ .$$

For an arbitrary $v \in \mathbb{R}^{m+n}$ it follows

$$(3.2) \qquad 2v^T M_k v = ((2-\omega)/\omega)\lambda_k v^T v + v^T C_k v \ .$$

From (2.2) we get further $M_k z_{k+1} = N_k z_k$ or

$$z_{k+1} = M_k^{-1}(M_k - C_k) z_k = z_k - M_k^{-1} C_k z_k \quad \text{and}$$

$$(3.3) \qquad M_k(z_k - z_{k+1}) =: M_k p_k = C_k z_k \ .$$

From the definition of the Raleigh quotient we have

$$(3.4) \qquad z_k^T(\lambda_k I - \hat{A}) z_k = z_k^T C_k z_k = 0 \ .$$

Using $\lambda_{k+1} = z_{k+1}^T \hat{A} z_{k+1} / z_{k+1}^T z_{k+1}$, (3.4), (3.3) and (3.2) with $v := p_k$ we get

$$
\begin{aligned}
(\lambda_{k+1}-\lambda_k) z_{k+1}^T z_{k+1} &= z_{k+1}^T \hat{A} z_{k+1} - \lambda_k z_{k+1}^T z_{k+1} = -z_{k+1}^T C_k z_{k+1} \\
&= -(z_k - p_k)^T C_k (z_k - p_k) \\
&= -z_k^T C_k z_k + p_k^T C_k z_k + z_k^T C_k p_k - p_k^T C_k p_k \\
&= 2p_k^T C_k z_k - p_k^T C_k p_k \\
&= 2p_k^T M_k p_k - p_k^T C_k p_k \\
&= \frac{2-\omega}{\omega} \lambda_k p_k^T p_k \ .
\end{aligned}
$$

□

The convergence of the sequences $\{\lambda_k\}$ and $\{z_k\}$ is now easily derived.

<u>*Lemma 2*</u>. *If z_o is chosen such that $\lambda_o > 0$, then for $0 < \omega < 2$, $\{\lambda_k\}_{k>o}$ is a monotonically increasing sequence which converges to an eigenvalue λ of $\hat{A}$. The sequence $\{z_k\}_{k>o}$ converges to an eigenvector z belonging to λ.*

<u>Proof</u>: From (3.1) it follows that $\{\lambda_k\}$ is monotonically increasing if $\lambda_o > 0$. Since $\lambda_k \le \rho(\hat{A})$ for $k \ge 0$, the sequence $\{\lambda_k\}$ is bounded and has a limit $\lambda > 0$. Again from (3.1) it follows that $p_k := z_k - z_{k+1}$ converges

to zero, i.e., $\{z_k\}$ has a limit z with $||z|| = 1$. Now from (3.3), the regularity of M_k for $\omega > 0$ and a continuity argument it follows that the limits $\lambda \neq 0$ and $z \neq 0$ satisfy the eigenvalue equation $C(\lambda)z = (\lambda I - \hat{A})z = 0$. □

4. Asymptotic convergence

Since the λ_k are Raleigh quotients of z_k and $\hat{A}$, the eigenvalue approximations λ_k converge much faster than the eigenvector approximations z_k (see [5] or [6]). Hence, for studying the asymptotic convergence behavior of the $\{z_k\}$, it is justified by Lemma 2 to consider $T(\omega,\lambda)$, where λ is an eigenvalue of $\hat{A}$.

Now, from (1.2) it follows that $C(\lambda) := \lambda I - A$ is a consistently ordered two-cyclic matrix ([8], p. 101). Thus the eigenvalues of $\hat{A}$ are $\pm\sigma_j$ $(j = 1,\dots,n)$ and 0. Hence the eigenvalues μ_j of the corresponding Jacobi iteration matrix $(1/\lambda)\hat{A}$ are

(4.1) $$\mu_j := \pm\sigma_j/\lambda \quad (j = 1,\dots,n)$$

and 0. Since by Lemma 2 there is $\lambda = \sigma_i$ for some index $i \in \{1,\dots,n\}$, we have $\mu_i = \pm 1$ for some $i \in \{1,\dots,n\}$. Further, between the eigenvalues τ_j of the SOR matrix $T(\omega,\lambda)$ and the eigenvalues μ_j of the Jacobi matrix $(1/\lambda)\hat{A}$, the well-known relation of Young holds ([8], p. 110)

(4.2) $$(\tau_j + \omega - 1)^2 = \tau_j\omega^2\mu_j^2 \quad (j = 1,\dots,n) .$$

For $\mu_j^2 = 1$ we get $\tau_{i,1} = 1$ and $\tau_{i,2} = (\omega - 1)^2$, i.e., $T(\omega,\lambda)$ has always an eigenvalue $\tau = 1$.

Now, as in [6], Theorem 3, we conclude for the convergence of the sequence $\{z_k\}$ it is necessary that

(4.3) $$q(\omega,\lambda) := \max\{|\tau_i| \mid \tau_i \text{ eigenvalue of } T(\omega,\lambda),\ \tau_i \neq 1\} < 1$$

and that no principal vector corresponds to the eigenvalue 1. By studying (4.2) it is easily seen that $q(\omega,\lambda) < 1$ iff

(4.4) $$\mu := \max\{|\mu_j| \mid \mu_j \text{ eigenvalue of } \hat{A}/\lambda,\ |\mu_j| \neq 1\} < 1.$$

But from (1.1) and (4.1) we conclude $\mu < 1$ iff $\lambda = \sigma_1 = ||A||_2$. Further, Schwarz [6] and Kolb [4] show that there exists no principal vector for the eigenvalue $\tau = 1$ (an extensive study of the principal vectors of $T(\omega,\lambda)$ is given in [4].

Thus, remembering Lemma 2, we have proven

Theorem 1. If z_o is chosen such that $\lambda_o > 0$, then, for $0 < \omega < 2$ the sequences $\{\lambda_k\}_{k\geq 0}$ and $\{z_k\}_{k\geq 0}$, generated by SOR, converge to $\sigma_1 = \|A\|_2$, resp. to $\hat{z} := (x,y)^T$, where x and y are left, resp. right, singular vectors of the given matrix A, corresponding to σ_1.

Remark 1: The assumption on z_o is easily satisfied: Choose $\tilde{z}_o = (x_o, y_o)^T$ such that $\tilde{\lambda}_o := 2x_o^T A y_o \neq 0$; if $\tilde{\lambda}_o < 0$, let $z_o = (-x_o, y_o)^T$, otherwise $z_o = \tilde{z}_o$.

The order of convergence is linear; we can try to minimize the convergence quotient $q(\omega,\sigma_1)$ by choosing an appropriate relaxation factor ω. With $\mu_j := \sigma_j/\sigma_1$ $(j = 1,\dots,n)$ and according to (4.4) $\mu := \max\{\mu_j \mid \mu_j \neq 1\}$ it follows ([6],[4]).

Theorem 2. The (asymptotic) convergence quotient $q(\omega,\sigma_1)$ becomes minimal for

$$(4.5) \qquad \omega = \omega_{opt} := \frac{2}{1+\sqrt{1-\mu^2}} \; ;$$

the minimal convergence quotient is

$$(4.6) \qquad q(\omega_{opt},\sigma_1) = \omega_{opt} - 1.$$

Remark 2: The convergence quotient of the vector iteration applied to $A^T A$ (i.e., SOR for $\omega = 1$) is $q(1,\sigma_1) = \mu^2$. Hence, if $\varepsilon^2 := 1 - \mu^2 \ll 1$, i.e., μ is very near to 1, SOR promises, at least asymptotically, a large increase in the rate of convergence, which can be measured by (see [8], p. 67)

$$(4.7) \qquad \frac{\ln(q(\omega_{opt},\sigma_1))}{\ln(q(1,\sigma_1))} \approx \frac{2}{\varepsilon} \, .$$

5. Global convergence.

We examined algorithm SOR with a lot of test matrices, where $\omega = \omega_o$ has been chosen from the beginning. The results were disappointing. The acceleration of convergence, indicated in Remark 2, couldn't be observed, but there was no systematic deviation.

Evidently the global convergence behavior depends on the choice of z_o. The following Theorem 3 partly explains this phenomenon.

At first we prove

Lemma 3. In the k-th step of algorithm SOR *let*

$$r_k := z_k - \hat{A} z_k/\lambda_k := (r_{k,1}, r_{k,2})^T \in \mathbb{R}^{m+n},$$

$$p_k := z_k - z_{k+1} \in \mathbb{R}^{m+n}, \quad v_k := Ay_k \in \mathbb{R}^m,$$

(5.1) $\alpha_k := 1 - 2\omega^2 v_k^T r_{k,1}/\lambda_k, \quad \pi_k := \| p_k \|^2$; then

(5.2) $$\frac{\lambda_{k+1}}{\lambda_k} = g_k(\omega) := \frac{\alpha_k + 2\pi_k/\omega}{\alpha_k + \pi_k}$$

holds, where both, α_k and π_k, depend on ω.

Proof: With (2.5)' and the definition of r_k it follows

(5.3) $$x_{k+1} = x_k - \omega r_{k,1}, \quad y_{k+1} = y_k - \omega r_{k,2} - \omega^2 A^T r_{k,1}/\lambda_k.$$

For $p_k := z_k - z_{k+1} =: (p_{k,1}, p_{k,2})^T$ we get

(5.4) $$p_{k,1} = \omega r_{k,1}; \quad p_{k,2} = \omega r_{k,2} + \omega^2 A^T r_{k,1}/\lambda_k.$$

Now, using $z_k^T z_k = 1$ and $\lambda_k = 2x_k^T A y_k$, we get

$$z_k^T p_k = \omega^2 v_k^T r_{k,1}/\lambda_k \quad \text{with} \quad v_k := Ay_k$$

and together with (5.3)

(5.5) $$\| z_{k+1} \|^2 = z_k^T z_k - 2 z_k^T p_k + p_k^T p_k = \alpha_k + \pi_k$$

with α_k and π_k according to (5.1). From Lemma 1 and (5.5) it follows

$$\frac{\lambda_{k+1}}{\lambda_k} = \frac{\| z_{k+1} \|^2 + (2/\omega - 1)\pi_k}{\| z_{k+1} \|^2}$$

$$= \frac{\alpha_k + 2\pi_k/\omega}{\alpha_k + \pi_k} =: g_k(\omega). \qquad \square$$

Theorem 3. Let $z_k =: (x_k, y_k)^T$, λ_k, λ_{k+1} be generated by SOR. *Then*

(5.6) $$\lambda_{k+1}/\lambda_k \leq 2/\omega ,$$

if either $\quad$ *a)* $0 < \omega \leq \sqrt{2}$

or $\quad$ *b)* $0 < \omega < 2,\ 0 < x_k^T x_k < \frac{2}{3}$ *is satisfied.*

<u>Proof</u>: From Lemma 3 it follows directly that (5.6) holds for $\alpha_k \geq 0$.

Because of $\lambda_k = 2x_k^T A y_k =: 2x_k^T v_k$ it holds

$$\lambda_k^2/4 = (x_k^T v_k)^2 \leq (x_k^T x_k)(v_k^T v_k). \tag{5.7}$$

From the definition of α_k and $r_{k,1}$ and with (5.7) it follows

$$\alpha_k = 1 - 2\omega^2 v_k^T r_{k,1}/\lambda_k$$

$$= 1 - 2\omega^2 v_k^T x_k/\lambda_k + 2\omega^2 v_k^T v_k/\lambda_k^2 \qquad \text{or}$$

$$\alpha_k \geq 1 - \omega^2 + \omega^2/(2x_k^T x_k). \tag{5.8}$$

Since $0 < x_k^T x_k < 1$ we conclude $\alpha_k \geq 1 - \omega^2/2 \geq 0$ for $0 \leq \omega \leq \sqrt{2}$.
If $0 < x_k^T x_k < 2/3$ then from (5.8)

$$\alpha_k \geq 1 - \omega^2/4 = (4-\omega)^2/4 > 0 \qquad \text{for } 0 < \omega < 2 . \qquad \square$$

Numerical examples indicate that (5.6) holds in most cases for $0 < \omega < 2$ without further restrictions. Now let us assume that (5.6) holds for $0 < \omega < 2$ and $k \geq 0$, which can be written in the form

$$\lambda_{k+1} \leq (2/\omega)^{k+1} \Lambda_o. \tag{5.9}$$

Then, if $\lambda_o << \sigma_1$ and ω_{opt} near 2, the increasing of the sequence $\{\lambda_k\}_{k\geq o}$ is bounded by (5.9) if from the beginning of the iteration $\omega = \omega_{opt}$ is chosen. This explains why the speed with which $\{\lambda_k\}_{k\geq o}$ approximates σ_1, depends on λ_o, i.e., on z_o. The question, how to choose ω at the first steps of SOR, and how to estimate ω_{opt} for the asymptotic optimal convergence is dealt with in the last section.

<u>6. Choice of the relaxation parameter</u>

Algorithm SOR is terminated, if either $\lambda_{k+1} - \lambda_k$ or $\|r_k\|$ is less than some bound ε, where the residium vector $r_k = (r_{k,1}, r_{k,2})^T$ is given by

$$(6.1)\qquad \begin{aligned} r_{k,1} &:= x_k - Ay_k/\lambda_k, \\ r_{k,2} &:= y_k - A^T x_k/\lambda_k. \end{aligned}$$

r_k can be computed without an extra matrix-vector multiplication, since $A^T x_k$ can be taken from step (2.6b)' of the preceding iteration, whereas Ay_k can be stored for step (2.6a)'.

As we have seen in the last section, it may not be appropriate to choose ω_{opt} from the beginning, even if ω_{opt} is known. Instead, it offers itself from Lemma 3 to choose $\omega = \omega_k$ such that for the function g_k in (5.2) the value ω_k with $g_k(\omega_k) = \max\{g_k(\omega) | 0 < \omega < 2\}$ is determined (it can be seen that $g_k(0) = g_k(2) = 1$). The corresponding strategy is realized in [4]. There are two difficulties: It isn't possible to determine ω_k analytically; to get an approximation ω_k numerically requires evaluation of g_k for different values of ω. Yet in g_k the new vector z_{k+1} is involved, i.e., for each evaluation, a SOR step has to be carried out. It is shown in [4], that g_k can be expressed as a rational function of ω, where the coefficients are scalar products of $r_{k,1}$, $r_{k,2}$ and $A^T r_{k,1}$. Thus, if the coefficients are computed once, the function g_k can be easily evaluated for different ω. Numerical examples show the following rule: There is only one maximum ω_k for $0 < \omega < 2$, and it holds that $\omega_k \leq \omega_{k+1} \leq \omega_{opt}$.

The second difficulty is that for about $k \geq 10$ the function g_k becomes so flat that ω_k can't be determined numerically. On the other side,this behavior of g_k indicates that λ_k is so near to σ_1 that the asymptotic formula for ω_{opt} can be used.

For estimating ω_{opt}, in principle all strategies known from linear systems can be applied (see, e.g.,[9], p. 209). One simple choice is: Let z_{k-2}, z_{k-1}, z_k be computed with the same value ω and assume $\omega \leq \omega_{opt}$. Then

$$(6.2)\qquad \frac{\| z_k - z_{k-1} \|}{\| z_{k-1} - z_{k-2} \|} \approx q(\omega, \sigma_1) =: \tau;$$

this value of τ inserted in (4.2) yields some estimate $\tilde{\mu}$ for μ and from (4.5) we get some estimate $\tilde{\omega}_{opt}$.

Algorithm SOR as it is described in Section 2, is nearly as simple as vector iteration, whereas the computation of an appropriate ω by maximizing the function g_k is comparitively involved. To keep the simplicity of the algorithm SOR the following "rule of thumb" is recommended: Choose $\omega_k := \tilde{\omega}$ for $k = 1,\dots,k_o$ and apply (6.2) for $k = k_o$; eventually (6.2) can be applied repeatedly; from a lot of numerical experiments the values $10 \leq k_o \leq 20$ (depending on the size of A) and $\tilde{\omega} = 1.25$ seem appropriate. If by accident, $\omega_{opt} < \tilde{\omega}$ which is indicated by an oscillation of $\|r_k\|$ then it is recommended to use $\tilde{\omega}$ further.

With these modifications SOR reveals itself as a very simple and robust algorithm for computing the spectral norm of an arbitrary $m \times n$-matrix A. It should be added that SOR is well suited for parallel computation.

References

[1] Forsythe,G.E., C.B. Moler: Computer solution of linear algebraic systems, Prentice Hall, Englewood Cliffs, 1967.

[2] Golub, G.H., Kahan, W.: Calculating the singular values and pseudoinverse of a matrix, J. SIAM Numer. Anal., Ser. B2, 205-224 (1965).

[3] Golub, G.H., C. Reinsch: Singular value decomposition and least squares solutions, in "Handbook for Automatic Computation", Vol. II Linear Algebra, 134-151, Springer, Berlin-Heidelberg-New York, 1971.

[4] Kolb, G.: Relaxationsmethoden zur Berechnung der Spektralnorm beliebiger Matrizen, Dissertation, Universität Mannheim, 1980.

[5] Ruhe, A.: SOR-methods for the eigenvalue problem with large sparse matrices, Math. of Comp. 28, 695-710 (1974).

[6] Schwarz, H.R.: The method of coordinate overrelaxation for (A-λB)x=0, Numer. Math. 23, 135-151 (1974).

[7] Schwarz, H.R.: The eigenvalue problem (A-λB)x=0 for symmetric matrices of high order, Comp. Math. in Appl. Mech. and Eng. 3, 11-28 (1974).

[8] Varga, R.S.: Matrix iterative analysis, Prentice Hall, Englewood Cliffs, 1962.

[9] Young, D.M.: Iterative solution of large linear systems, Academic Press, New York-San Francisco-London, 1971.

NUMERICAL COMPUTATION OF PERIODIC SOLUTIONS OF A NONLINEAR WAVE EQUATION

Theodor Meis and Wolfgang Baaske
Mathematisches Institut
der Universität zu Köln

Nonlinear wave equations of the form

$$u_{tt}(x,t) = u_{xx}(x,t) + \gamma f(u(x,t))$$

with a fixed $\gamma \in \mathbb{R}$ and a given function $f:\mathbb{R} \to \mathbb{R}$ are of great importance in diverse fields of physical science. Above all, one is interested in global solutions in $\mathbb{R}^2$, especially in solutions that are periodic in x and t. Examples can be found in A.H. Nayfeh's and D.T. Mook's book [2]. According to the application, the functions f may be entirely different. We experimented with the following mappings:

$$f_1(z) = z^3 ,$$

$$f_2(z) = z - \tfrac{1}{6}z^3 ,$$

$$f_3(z) = \sin(z) .$$

In principle, our methods are applicable to many other functions, provided that they are real-analytic and odd.

Consider the *boundary value problem A:*

$$\begin{aligned}
&u_{tt}(x,t) = u_{xx}(x,t) + \gamma f(u(x,t)) && (x,t \in \mathbb{R})\\
&u(-\pi/2,t) = u(+\pi/2,t) = 0 && (t \in \mathbb{R})\\
&u(x,0) = u(x,\tau) && (x \in \mathbb{R})\\
&u_t(x,0) = u_t(x,\tau) && (x \in \mathbb{R}) .
\end{aligned}$$

We can prescribe γ and f and determine τ and u. Sometimes it is better to specify τ and f. Then the unknowns are γ and u. We take the case $f(z) = z^3$ as a model problem. For the present, we confine ourselves to this special case.

Let $\alpha \in \mathbb{R}$, $\alpha \neq 0$ and let the pair (τ,u) be a solution of A. Then, the pair $(\tau,\alpha u)$ satisfies the same problem with γ/α^2 instead of γ.

Furthermore, we consider the transformation from u into $\hat{u}$ with $\hat{u}(x,t) = u(\nu x, \nu t)$ and an odd integer ν. The pair $(\tau, \hat{u})$ solves the problem A with $\gamma \cdot \nu^2$ replacing γ. These transformations suggest to add the following normalizing conditions to the boundary value problem: $u(0,0) = 1$, u has no period $< 2\pi$ in x.

For simplicity, we define $\omega = 2\pi/\tau$, $s = \omega t$ and $\hat{u}(x,s) = u(x,t)$. This yields the differential equation

$$\omega^2 \hat{u}_{ss}(x,s) = \hat{u}_{xx}(x,s) + \gamma f(\hat{u}(x,s)) .$$

In the following, we again write u and t instead of $\hat{u}$ and s.

Boundary value problem B:

$$\begin{aligned}
&\omega^2 u_{tt}(x,t) = u_{xx}(x,t) + \gamma f(u(x,t)) && (x,t \in \mathbb{R})\\
&u(-\pi/2,t) = u(+\pi/2,t) = 0 && (t \in \mathbb{R})\\
&u(x,0) = u(x,2\pi) && (x \in \mathbb{R})\\
&u_t(x,0) = u_t(x,2\pi) && (x \in \mathbb{R})\\
&u(0,0) = 1 &&\\
&u \text{ has no period } < 2\pi \text{ in } x. &&
\end{aligned}$$

1. Solutions with $|\gamma| \ll 1$

For $\gamma = 0$ the problem is linear. The solutions are

$$\tau = 2\pi$$

$$u(x,t) = \sum_{\nu=0}^{\infty} \alpha_\nu \cos((2\nu+1)x)\cos((2\nu+1)t)$$

with

$$\sum_{\nu=0}^{\infty} \alpha_\nu = 1 \quad \text{and} \quad \alpha_0 \neq 0 .$$

The coefficients α_ν have to be choosen in such a way, that the series converges to a twice continously differentiable function. For $\gamma \neq 0$, $|\gamma| \ll 1$, B can be reduced to a two-point boundary value problem.

Consider

$$X_1 = \mathcal{L}^2((0,2\pi))$$

$$X_2 = \mathcal{L}^2(\Omega) \quad , \ \Omega = (0,2\pi) \times (0,2\pi)$$

$$X_1 \supset V_1 = \overline{\text{span}\{\cos((2\mu-1)x) \mid \mu \in \mathbb{N}\}}$$

$$X_2 \supset V_2 = \overline{\text{span}\{\cos((2\mu-1)x)\cos((2\mu-1)t) \mid \mu \in \mathbb{N}\}}$$

$$X_2 \supset W_2 = \overline{\text{span}\{\cos((2\mu-1)x)\cos((2\nu-1)t) \mid \mu,\nu \in \mathbb{N},\ \mu \neq \nu\}} .$$

The following projections belong to these subspaces

$$P_V : X_2 \to V_2$$

$$P_W : X_2 \to W_2 .$$

The prolongation

$$J : X_1 \to X_2$$

is defined by

$$\cos((2\mu-1)x) \to \cos((2\mu-1)x)\cos((2\mu-1)t) .$$

We make the assumption that there exist $\varepsilon > 0$ and functions $u,v,w : \mathbb{R}^2 \times [-\varepsilon,\varepsilon] \to \mathbb{R}$, $g : [-\varepsilon,\varepsilon] \to \mathbb{R}$ with the following properties:

(1) for constant γ, $u(x,t,\gamma)$ is a solution of the boundary value problem B with $\omega = 1 - \gamma g(\gamma)$,

(2) $u = v + \gamma w$; for constant γ it holds $v \in V_2$, $w \in W_2$,

(3) v,w and all partial derivatives of first and second order with respect to x and t are continuous in $\overline{\Omega} \times [-\varepsilon,\varepsilon]$; g is continuous.

This leads to

(4) $\omega^2 v_{tt} - v_{xx} = -2\gamma g(\gamma) v_{xx} + O(\gamma^2)$.

(5) $\omega^2 w_{tt} - w_{xx} = D(w) + O(|\gamma|)$ with

$$D(\cos((2\mu-1)x)\cos((2\nu-1)t))$$
$$= [4(\mu-\nu)(\mu+\nu-1)]\cos((2\mu-1)x)\cos((2\nu-1)t)$$

(6) $u^3 = P_V(v^3) + P_W(v^3) + O(|\gamma|)$.

Decomposing the differential equation into the parts relating to V_2 and W_2 yields for $\gamma = 0$:

(7) $D(\tilde{w}) = P_W(\tilde{v}^3)$

(8) $2\delta\tilde{v}_{xx} + P_V(\tilde{v}^3) = 0$

with $\tilde{v}(x,t) = v(x,t,0)$, $\tilde{w}(x,t) = w(x,t,0)$, $\delta = g(0)$.

$D^{-1} : W_2 \to W_2$ is a bounded linear mapping. It is $\| D^{-1} \| = \frac{1}{8}$, because for $\mu, \nu \in \mathbb{N}$ and $\mu \neq \nu$

$$D^{-1}(\cos((2\mu-1)x)\cos((2\nu-1)t))$$

$$= \frac{1}{4(\mu-\nu)(\mu+\nu-1)} \cos((2\mu-1)x)\cos((2\nu-1)t)$$

and

$$\left| \frac{1}{4(\mu-\nu)(\mu+\nu-1)} \right| \leq \frac{1}{8} \frac{1}{\min(\mu,\nu)} \leq \frac{1}{8} .$$

Equality holds for $\mu = 2$, $\nu = 1$ and $\mu = 1$, $\nu = 2$. Moreover

$$D^{-1}(W_2 \cap C^2(\bar{\Omega})) \subset W_2 \cap C^2(\bar{\Omega}) .$$

Thus, equation (7) permits the calculation of $\tilde{w}$ from $\tilde{v}$.

Equation (8) is an equation for $\tilde{v}$ and δ. Since J is a bijective mapping, there exist a $\hat{v} \in V_1$ with $\tilde{v} = J(\hat{v})$ and $\tilde{v}_{xx} = J(\hat{v}'')$. The question, if there are two functions u and g for which our assumption holds, leads to the question: Do $\hat{v} \in V_1$ and $\delta \in \mathbb{R}$ exist which solve equation (9)?

$$(9) \quad 2\delta\hat{v}'' + J^{-1}(P_V((J\hat{v})^3)) = 0 \qquad , \hat{v}(0) = 1 .$$

Our numerical investigations show that this eigenvalue-problem has at least one isolated solution:

$$\hat{v}(x) \approx 0.9855867\cos(x) + 0.142085 \cdot 10^{-1}\cos(3x)$$
$$+ 0.2020 \cdot 10^{-3}\cos(5x) + 0.29 \cdot 10^{-5}\cos(7x)$$

$$\delta \approx 0.27459 .$$

2. *Application of difference methods*

Difference methods that are primarily developed for solving initial value problems can also be used for the computation of periodic solutions. At first, we transform the differential equation (boundary value problem A) into a first order system

$$w_t = \begin{pmatrix} 0 & 0 & 0 \\ 0 & 0 & 1 \\ 0 & 1 & 0 \end{pmatrix} w_x + \begin{pmatrix} w_3 \\ 0 \\ \gamma w_1 \end{pmatrix} + \eta w_{xx}$$

with $w = (u,u_x,u_t)^T = (w_1,w_2,w_3)^T$.

The term ηw_{xx}, $\eta > 0$, is an additional viscosity term. For this reason, the new system is parabolic. The difference schemes for solving hyperbolic systems automatically add such viscosity terms. Our explicit specification in the differential equation has the advantage that we exactly know the size of the viscosity.

In consequence of this term, a periodic solution $(w(x,\tau) = w(x,0))$ is impossible. Instead of this, we get $w(x,\tau) = \rho w(x,0)$ with $\rho < 1$, but $\lim_{\eta \to 0} \rho(\eta) = 1$.

We obtained the best efficiency with the simplest explicit difference scheme:

$$w(x,t+h) = w(x,t) + \frac{h}{2\Delta x} \begin{pmatrix} 0 & 0 & 0 \\ 0 & 0 & 1 \\ 0 & 1 & 0 \end{pmatrix} [w(x+\Delta x,t) - w(x-\Delta x,t)]$$

$$+ \frac{\eta h}{(\Delta x)^2} [w(x+\Delta x,t) + w(x-\Delta x,t) - 2w(x,t)].$$

For $\gamma = 0$ it is stable if $\Delta x \leq 2\eta$ and $2\eta h \leq (\Delta x)^2$. In this case, the difference equations can even be solved exactly by seperation (c.f. [1]). For $\gamma \neq 0$, we use the following iteration method:

(1) Start with $\gamma > 0$ and with an initial guess $\tilde{w}(x,0)$, where $\tilde{w}_1(0,0) = 1$.

(2) Solve the initial value problem up to a level $\tilde{\tau} = nh$, where $\tilde{w}_3(0,t)$ changes the sign for the second time.

(3) Replace $\tilde{w}(x,0)$ by $\tilde{w}(x,\tilde{\tau})/w_1(0,\tilde{\tau})$ and continue with (2).

We found that $\tilde{\tau}$ and $\tilde{w}$ converge with this power method. Denote

$$\tau = \lim \tilde{\tau}$$

$$w = \lim \tilde{w}$$

$$w(x,\tau) = \rho w(x,0) \qquad , \ x \in \mathbb{R}.$$

ρ and τ depend on η, h and Δx. If all these values are small enough, ρ is close to 1. Then, τ is approximately a polynomial of first degree in ρ. Therefore, one can calculate a limit of τ for $h = \eta = \Delta x = 0$. All details are described by W. Baaske [1]. For $0 < \gamma \leq 1$, i.e. $2\pi < \tau < 10$, the results obtained by this procedure only differ by 1% from the best results that we computed with trigonometric approximations. The represented iteration method can certainly be improved. We did not follow this way, because the trigonometric expansions promised to be more efficient.

3. *Simple trigonometric approximations*

Nayfeh and Mook [2] propose the function $u = \cos(x)\cos(t)$ as a first approximation to the solution of the boundary value problem B. It follows

$$\omega^2 u_{tt} - u_{xx} - \gamma u^3 = (-\omega^2 + 1 - \tfrac{9}{16}\gamma)\cos(x)\cos(t)$$
$$+ \tfrac{3}{16}\gamma\cos(3x)\cos(t) + \tfrac{3}{16}\gamma\cos(x)\cos(3t) + \tfrac{1}{16}\gamma\cos(3x)\cos(3t).$$

The first coefficient vanishes if $16\omega^2 + 9\gamma = 16$ or $64\pi^2 + 9\gamma\tau^2 = 16\tau^2$. Points on this curve are for example

τ	γ	$\gamma_{trig.}$
6	-0.172	-0.176
7	0.345	0.353
8	0.681	0.692
10	1.076	1.075

$\gamma_{trig.}$ is our best value (c.f. table 1).

Now, we want to investigate a somewhat better approximation. In this context, we use the abbreviation

$$z_{jk}(x,t) = \cos((2j-1)x)\cos((2k-1)t)$$

and the addition formula

$$4\cos((2l-1)\alpha)\cos((2m-1)\alpha)\cos((2n-1)\alpha)$$
$$= \cos((2l+2m+2n-3)\alpha) + \cos((2l+2m-2n-1)\alpha)$$
$$+ \cos((2l-2m+2n-1)\alpha) + \cos((-2l+2m+2n-1)\alpha) .$$

Our approximation has the form

$$u = a[z_{11} + bz_{21} + cz_{12} + dz_{22}]$$

with coefficients $a,b,c,d \in \mathbb{R}$.
We immediately see

$$\frac{\omega^2 u_{tt} - u_{xx}}{a} = (-\omega^2 + 1)z_{11} + (-\omega^2 + 9)bz_{21} + (-9\omega^2 + 1)cz_{12} + (-9\omega^2 + 9)dz_{22} .$$

The calculation of u^3 is more complicated.

$$\begin{aligned}\frac{16u^3}{a^3} = {} & (9 + 9b + 9c + 3d + 18b^2 + 18c^2 + 12d^2 \\ & + 6bc + 12bd + 12cd + 6bc^2 + 6b^2c + 24bcd)z_{11} \\ & + (3 + 18b + 3c + 6d + 6c^2 + 12bc + 24cd + 9b^3 \\ & + 12bc^2 + 9b^2d + 18bd^2)z_{21} \\ & + (3 + 3b + 18c + 6d + 6b^2 + 12bc + 24bd + 9c^3 \\ & + 12b^2c + 9c^2d + 18cd^2)z_{12} \\ & + (1 + 6b + 6c + 12d + 24bc + 3b^3 + 3c^3 \\ & + 18b^2d + 18c^2d + 9d^3)z_{22} + \dots\end{aligned}$$

The coefficients of z_{11}, z_{21}, z_{12}, z_{22} must vanish in the defect $\omega^2 u_{tt} - u_{xx} - \gamma u^3$. Together with the condition

$$u(0,0) = a(1 + b + c + d) = 1$$

one obtains five equations for a,b,c,d,γ,ω. If one prescribes γ in a neighbourhood of 0 (or ω in a neighbourhood of 1) it is easy to compute the other five quantities with Newton's method. One can take

$$a = 1,\ b = c = d = 0,\ \omega = 1$$

or

$$a = 1,\ b = c = d = 0,\ \gamma = 0$$

as initial guess.

4. More general trigonometric approximations

The approach which is used in the previous section can easily be generalized. With

$$u = \sum_{j=1}^{n} \sum_{k=1}^{n} c_{jk} z_{jk} ,$$

it follows

$$u^3 = \sum_{j=1}^{3n-1} \sum_{k=1}^{3n-1} d_{jk} z_{jk} .$$

The coefficients d_{jk} are polynomials in the coefficients c_{jk}. The defect of the differential equation becomes

$$\omega^2 u_{tt} - u_{xx} - \gamma u^3 = \sum_{j=1}^{3n-1} \sum_{k=1}^{3n-1} e_{jk} z_{jk} .$$

In order to get an approximate solution of the differential equation we postulate for $j,k = 1(1)n$

$$e_{jk} = [-\omega^2 (2k-1)^2 + (2j-1)^2] c_{jk} - \gamma d_{jk} = 0 .$$

Additionally, we have to notice the normalization

$$u(0,0) = \sum_{j=1}^{n} \sum_{k=1}^{n} c_{jk} = 1 .$$

Given $\tau = 2\pi/\omega$, one gets a nonlinear system of equations to determine the unknowns γ and c_{jk}. This system can be solved, for example, by using a secant method. Of course, it is possible to prescribe γ and to compute ω. Both methods have their own advantages.

From the practical point of view, the main question is how to calculate the d_{jk} as functions of the c_{jk}.

Our method is based on discrete orthogonality relations for the functions z_{jk}. We want to explain these relations. Therefore, with

$$N = 6n - 1, \quad m = 3n - 1, \quad h = 2\pi/N$$

$$\sigma_j = \begin{cases} 1/2 & \text{for } j = 0 \\ 1 & \text{otherwise} \end{cases} ,$$

we define

$$\langle v,w \rangle = \frac{16}{N^2} \sum_{\alpha=0}^{m} \sum_{\beta=0}^{m} \sigma_\alpha \sigma_\beta v(\alpha h, \beta h) w(\alpha h, \beta h)$$

for $v,w : \bar{\Omega} \to \mathbb{R}$.

Theorem: For $j,k,\mu,\nu = 1(1)m$ it holds

$$\langle z_{jk}, z_{\mu\nu} \rangle = \delta_{j\mu} \delta_{k\nu} \qquad (\delta_{j\mu} = \text{Kronecker symbol}).$$

This immediately leads to

$$d_{jk} = \langle u^3, z_{jk} \rangle .$$

The pointwise calculation of u^3 causes no difficulties. The computation effort of a straightforward algorithm is of order $O(n^4)$. We realized the algorithm by numerous examples for $n = 2(1)5$.

Proof: Let $i = \sqrt{-1}$, $\varepsilon_j = 2j-1$. It is well-known for $j,k = 1(1)N-1$ that

$$\frac{1}{N} \sum_{\alpha=0}^{N-1} \exp(ij\alpha h)\exp(-ik\alpha h) = \delta_{jk}$$

or

$$\frac{1}{N} \sum_{\alpha=-m}^{m} \exp(ij\alpha h)\exp(-ik\alpha h) = \delta_{jk} .$$

The functions

$$z_{jk}(x,t) = \cos(\varepsilon_j x)\cos(\varepsilon_k t)$$

are even functions in x and t. Therefore it holds

$$\langle z_{jk}, z_{\mu\nu} \rangle = \frac{4}{N^2} \sum_{\alpha=-m}^{m} \sum_{\beta=-m}^{m} \cos(\varepsilon_j \alpha h)\cos(\varepsilon_k \beta h)\cos(\varepsilon_\mu \alpha h)\cos(\varepsilon_\nu \beta h)$$

$$= \frac{4}{N^2} \sum_{\alpha=-m}^{m} [\cos(\varepsilon_j \alpha h)\cos(\varepsilon_\mu \alpha h) \sum_{\beta=-m}^{m} \cos(\varepsilon_k \beta h)\cos(\varepsilon_\nu \beta h)]$$

$$= [\frac{2}{N} \sum_{\alpha=-m}^{m} \cos(\varepsilon_j \alpha h)\cos(\varepsilon_\mu \alpha h)][\frac{2}{N} \sum_{\alpha=-m}^{m} \cos(\varepsilon_k \alpha h)\cos(\varepsilon_\nu \alpha h)].$$

Both factors show the same structure. Thus, it is sufficient to evaluate the first one. With

$$\exp(i\alpha Nh) = \exp(i\alpha 2\pi) = 1$$

it follows

$$\begin{aligned} &4\cos(\varepsilon_j \alpha h)\cos(\varepsilon_\mu \alpha h) \\ &\quad = \exp(i\varepsilon_j \alpha h)\exp(-i(N-\varepsilon_\mu)\alpha h) \\ &\quad + \exp(i\varepsilon_j \alpha h)\exp(-i\varepsilon_\mu \alpha h) \\ &\quad + \exp(i(N-\varepsilon_j)\alpha h)\exp(-i(N-\varepsilon_\mu)\alpha h) \\ &\quad + \exp(i(N-\varepsilon_j)\alpha h)\exp(-i\varepsilon_\mu \alpha h) . \end{aligned}$$

All the numbers ε_j, ε_μ, $N-\varepsilon_j$, $N-\varepsilon_\mu$ lie in the intervall $(1,N-1)$. ε_j

τ	γ_2	γ_3	γ_4	γ_5	c_{11}	s
1.00	-46.58	-58.99	-52.9538	-55.9766	1.180123	.656
2.00	-12.77	-13.82	-13.6678	-13.9600	1.114256	.345
3.00	-5.546	-5.606	-5.58346	-5.58598	1.064816	.187
4.00	-2.570	-2.609	-2.60796	-2.60811	1.009705	.103
5.00	-1.044	-1.047	-1.04656	-1.04657	0.992542	.054
6.00	-.1758	-.1759	-.175884	-.175884	0.985875	.022
7.00	0.3529	0.3532	0.353221	0.353221	0.987162	.033
8.00	0.6899	0.6918	0.691792	0.691793	0.993346	.057
9.00	0.9107	0.9171	0.917063	0.917075	1.002129	.080
10.00	1.0574	1.0757	1.075317	1.075399	1.010522	.107
10.50	1.1111	1.1478	1.147197	1.147488	1.010188	.126
10.60	1.1205	1.1652	1.164597	1.165000	1.008364	.131
10.70	1.1296	1.1861	1.185817	1.186407	1.004808	.139
10.80	1.1384	1.2153	1.215748	1.216685	0.997539	.151
10.90	1.1467	1.2674	1.271510	1.273203	0.979809	.175
10.98	1.1532	1.3785	1.401704	1.404746	0.932961	.229
11.00	1.1548	1.4462	1.492028	1.495864	0.899907	.270

Table 1

τ	γ_2	γ_3	γ_4	γ_5	c_{11}	s
11.00	1.1548	-	0.329406	0.333239	1.729426	.980
11.06	1.1594			0.684479	1.301289	.455
11.10	1.1625	0.8186	0.871531	0.869859	1.179588	.304
11.20	1.1698	1.0393	1.052725	1.049506	1.090789	.201
11.30	1.1769	1.1028	1.108213	1.105018	1.069849	.177
11.40	1.1837	1.1341	1.136766	1.132589	1.062100	.169
11.50	1.1902	1.1543	1.155713	1.143697	1.061790	.169
11.60	1.1964	1.1695	1.170095	1.176956	1.051181	.164
11.70	1.2023	1.1819	1.181891	1.184217	1.052321	.162
11.80	1.2080	1.1924	1.192040	1.193378	1.052467	.162
11.90	1.2135	1.2018	1.201042	1.201975	1.052751	.163
12.00	1.2187	1.2103	1.209192	1.209918	1.053219	.164
13.00	1.2595	1.2708	1.266330	1.266858	1.062895	.183
14.00	1.2844	1.3095	1.299243	1.300223	1.075787	.207
15.00	1.2983	1.3368	1.309052	1.310642	1.093260	.239
16.00	1.3045	1.3567	1.452215	1.437628	1.059769	.276
17.00	1.3056	1.3718	1.382514	1.385691	1.095314	.280

Table 2

and ε_μ are odd, $N-\varepsilon_j$ and $N-\varepsilon_\mu$ are even. So $\varepsilon_j \neq N-\varepsilon_\mu$ and $\varepsilon_\mu \neq N-\varepsilon_j$ is always true. The above orthogonality relation for the function exp yields

$$\frac{2}{N} \sum_{\alpha=-m}^{m} \cos(\varepsilon_j \alpha h)\cos(\varepsilon_\mu \alpha h) = \delta_{j\mu} .$$

5. Numerical results

In the sections 3 and 4 we discussed two approximation methods to solve the boundary value problem B. The first algorithm is a special case of the second one for $n = 2$. Therefore it is sufficient to explain the results for the more general approach.

First, we consider $f(u) = u^3$. Table 1 and table 2 contain the figures for periods τ in the intervall (1,17). γ_n is an approximation for γ using the expansion with n^2 coefficients. To give an impression of the characteristic behaviour of the solution the coefficient c_{11} and

$$s = \sum_{j=1}^{5} \sum_{k=1}^{5} |c_{jk}| - |c_{11}|$$

are additionally tabulated for $n = 5$. For $\tau \in [4,9]$ the differences in γ_n are little, c_{11} is close to 1 and s is very small. A similar behaviour can be obtained for $\tau \in [12,17]$. On the other hand, the intervall [10,12] is of particular interest. As the figure 1 shows, there are qualitative differences in the shape of γ_2 and γ_5.

For $f(u) = \sin(u)$, $n = 2(1)5$ and $\tau \in [5,56]$, γ_n is a monotone function of τ without any peculiarity.

τ	γ_4	γ_5	c_{11}	s
5	-0.63688	-0.63688	1.000	.005
7	0.21371	0.21371	1.001	.003
9	0.56365	0.56365	0.999	.008
15	0.90544	0.90544	0.985	.024
23	1.01318	1.01318	0.952	.058
39	1.05793	1.05787	0.847	.163
56	1.06498	1.06446	0.714	.297

Table 3

For $f(u) = u - u^3/6$ we observe some differences

τ	γ_4	γ_5	c_{11}	s
5	-0.63914	-0.63914	1.000	.006
7	0.21447	0.21447	1.001	.003
9	0.56564	0.56564	0.999	.008
15	0.90849	0.90849	0.984	.026

Table 4

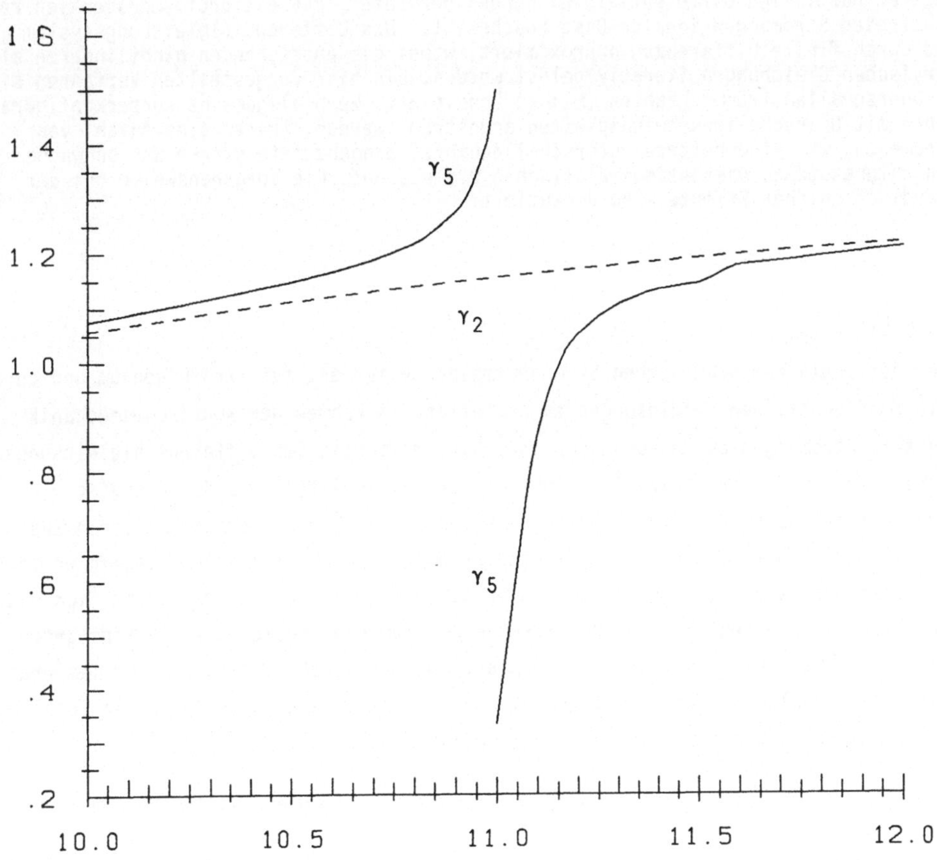

Figure 1

References

[1] W. Baaske: Periodische Lösungen einer nichtlinearen Wellengleichung, Diplomarbeit, Universität zu Köln (1981).

[2] A.H. Nayfeh, D.T. Mook: Nonlinear Oscillations, John Wiley & Sons, Inc., New York (1979).

Erfahrungen bei der Anwendung numerischer Verfahren zur Lösung nichtlinearer hyperbolischer Differentialgleichungssysteme

C. Weiland
Messerschmitt-Bölkow-Blohm GmbH
Ottobrunn

Zusammenfassung

In Form eines Überblicks wird von der numerischen Integration des Systems von quasilinearen partiellen Differentialgleichungen berichtet, das die dreidimensionalen reibungsfreien Strömungen idealer Gase beschreibt. Das Differentialgleichungssystem wird durch finite Differenzen approximiert, wobei die entstehenden nichtlinearen algebraischen Gleichungen iterativ gelöst werden. Die hier vorgestellten Verfahren sind für Überschallanströmmachzahlen ($M_\infty > 1$) konzipiert. Wenn allgemeine Körperkonfigurationen mit Überschallgeschwindigkeiten angeströmt werden, treten eine Anzahl von Phänomenen, wie eingebettete Unterschallgebiete, eingebettete starke und schwache Verdichtungsstöße, Diskontinuitätsflächen u.s.w., auf. Die Vorgehensweise bei der Behandlung solcher Effekte wird diskutiert.

I. Einleitung

Eines der Ziele der numerischen Strömungsmechanik ist es, für Strömungen um und durch reale Konfigurationen Feldlösungen zu erstellen. Im Rahmen der Kontinuumsmechanik sind alle Strömungen durch Lösungen eines Systems partieller Differentialgleichungen, welches aus den Navier-Stokes Gleichungen, der Kontinuitäts- und der Energiegleichung besteht, gegeben. Hinzu kommt eine Gleichung, die den thermodynamischen Zustand festlegt. Für den Fall der reibungsfreien Strömung ohne Wärmeleitung reduzieren sich die Navier-Stokes Gleichungen zu den Euler-Gleichungen. Ebenso vereinfacht sich die Energiegleichung erheblich. Sei das Strömungsmedium ein ideales Gas mit konstanten spezifischen Wärmen, so ist der Zusammenhang zwichen Druck p, Dichte ρ und Temperatur T durch die ideale Gasgleichung gegeben, und die innere Energie e sowie die Enthalpie h sind einfache lineare Funktionen der Temperatur.

Damit erhalten wir ein System gekoppelter partieller Differentialgleichungen 1. Ordnung, welches quasilinear ist.

Für zeitabhängige (instationäre) Strömungen ist dieses Differentialgleichungssystem grundsätzlich hyperbolisch. Für zeitunabhängige (stationäre) Strömungen ist das System dann hyperbolisch, wenn die Strömungsgeschwindigkeit $|\vec{v}|$ in jedem Punkt des Strömungsfeldes größer als die lokale Schallgeschwindigkeit c ist.

Dieser Bericht beschäftigt sich mit numerischen Lösungen des oben genannten Differentialgleichungssystems, in den Fällen, in denen es hyperbolisch ist. Es werden Berechnungsbeispiele für Strömungen um stumpfe Körper mit sehr schallnahen Anströmmachzahlen gegeben, wobei es wegen des Unterschallgebietes notwendig ist, die instationären

Bewegungsgleichungen zu integrieren und bei stationären Randbedingungen die zeitlich asymptotische Lösung als die gesuchte Lösung zu betrachten. In diesem Fall spricht man von einer "time-marching-procedure". Weiterhin wird ein Beispiel der Berechnung eines reinen Oberschallströmungsfeldes um einen Rotationskörper mit eingebettetem Verdichtungsstoß diskutiert. Hierzu genügt es, die stationären Gleichungen mit einer "space-marching-procedure" zu integrieren. Die Grundlagen dieser Verfahren, was die Formulierungen der Gleichungen, die Randbedingungen, die Koordinatentransformationen und deren diskrete Analoga betrifft, sind in den Arbeiten [1] bis [3] enthalten, wobei die verwendeten finiten Differenzenverfahren auf die Arbeiten [4] und [5] zurückgehen. Von besonderer Schwierigkeit ist die Berechnung von Strömungsfeldern um Flügel und Flügelrumpf-Kombinationen.

Für Strömungen um einen Deltaflügel werden die Bestimmungsgleichungen und speziell die verwendeten Koordinatentransformationen angegeben und Ergebnisse werden diskutiert.

II. Die Bestimmungsgleichungen

Das Differentialgleichungssystem für dreidimensionale stationäre Strömungen in symbolischer Schreibweise lautet:

$$\begin{aligned} \operatorname{div}(\rho\vec{v}) &= 0 && \text{Kontinuitätssatz} \\ \rho(\vec{v}\cdot\operatorname{grad})\vec{v} + \operatorname{grad}p &= 0 && \text{Impulssatz} \\ \vec{v}\cdot(\operatorname{grad}p - c^2\operatorname{grad}\rho) &= 0 && \text{Energiesatz} \end{aligned} \tag{1}$$

Es bedeuten ρ die Dichte, p der Druck, c die lokale Schallgeschwindigkeit und $\vec{v}$ der Geschwindigkeitsvektor. Das System (1) läßt sich konservativ formulieren und nimmt in zylindrischen Koordinaten z, r, φ die folgende Form an:

$$\frac{\partial E}{\partial z} + \frac{\partial F}{\partial r} + \frac{\partial G}{\partial \varphi} + H = 0 \tag{2}$$

E, F, G, H sind Vektoren, die sich aus den abhängigen Variablen ρ, ρu, ρv, ρw, e aufbauen (siehe [2]). Es sind u, v, w die Komponenten des Geschwindigkeitsvektors $\vec{v}$ in Richtung der zylindrischen Koordinaten und e bedeutet die Gesamtenergie mit

$$e = \frac{p}{\kappa-1} + \frac{(\rho u)^2 + (\rho v)^2 + (\rho w)^2}{2\rho}$$

Das konservative System (2) hat im Gegensatz zu einem nicht-konservativen System die Eigenschaft, als schwache Lösung die Stoß-Gleichungen (Rankine-Hugoniot-Gleichungen) zu enthalten [6]. Aus lösungstechnischen Gründen [2] muß eine quasi-konservative

Formulierung der Bewegungsgleichungen verwendet werden, die aus (2) durch Bildung der Jacobi-Matrizen von E, F, G bezüglich des Lösungsvektors $U^T = \{\rho, \rho u, \rho v, \rho w, e\}$ entsteht.

$$J(U)\,\frac{\partial U}{\partial z} + K(U)\,\frac{\partial U}{\partial r} + L(U)\,\frac{\partial U}{\partial \varphi} + H(U) = 0 \qquad (3)$$

Die Jacobi-Matrizen J(U), K(U), L(U) sind in [2] gegeben.

Für ein eindimensionales instationäres Testbeispiel wird in [7] nachgewiesen, daß das quasi-konservative Differentialgleichungssystem dieselben Konservativitätseigenschaften besitzt, wie das konservative und daß die diskreten Approximationen dieses Gleichungssystems konservativ bis auf Terme der $O(\Delta t^3)$ sind. Zur Veranschaulichung zeigt Bild 1 den Verlauf des Druckes, wie er sich bei der numerischen Lösung der eindimensionalen instationären Euler-Gleichungen ergibt, wenn diese konservativ, quasikonservativ bzw. nicht-konservativ formuliert werden. Es handelt sich dabei um eine von links kommende Verdichtungswelle, deren zeitliche Entwicklung nach 25, 99 und 300 Zeitschritten aufgetragen worden ist. Die für dieses Problem bekannte analytische Lösung ist zum Vergleich ebenfalls eingetragen. Während die numerische Approximation der nicht-konservativen Euler-Gleichungen offenbar einer ganz anderen Lösung zustrebt (a), geben sowohl die konservative (b), als auch die quasi-konservative (c) Formulierung das Druckprofil sehr gut wieder. Zahlreiche Genauigkeitstests, wie sie in [7] gemacht wurden, bestätigen dies.

III. Die Rechenkoordinatensysteme

Wie schon erwähnt, wurden zylindrische Koordinaten z, r, φ als Ausgangskoordinaten verwendet. Wir wählen als Rechenkoordinaten $\tilde{z}$, ξ, ϑ mit der Maßgabe, daß dadurch der physikalische Raum, der zwischen der Körperkonturfläche und der Verdichtungsstoßkonturfläche entsteht, auf einen mit ebenen Flächen berandeten Rechenraum abgebildet wird. Dabei sollen die Flächen der Körperkontur und der Verdichtungsstoßkontur Koordinatenflächen ξ = konstant sein. Als weiteres wollen wir fordern, daß in Ebenen z = konstant der Ursprung eines zwischengeschalteten lokalen Polarkoordinatensystems durch zwei frei wählbare Funktionen x_0, y_0 dargestellt werden kann. Dadurch wird gute Auflösung in Bereichen starker Körperkrümmung erreicht, in denen in aller Regel die größten Gradienten der Strömungsvariablen auftreten.
Insgesamt ergibt sich (Bild 2):

$$(z, r, \varphi) \rightarrow (\bar{z}, \bar{x}, \bar{y}) \rightarrow (\bar{z}, R, \vartheta) \rightarrow (\bar{z}, \xi, \vartheta)$$

mit

$$
\begin{aligned}
z &= \bar{z} \\
r &= (\bar{x}^2 + \bar{y}^2)^{1/2} \\
\varphi &= \arctan \frac{\bar{y}}{\bar{x}} \\
\bar{x}\,(\bar{z}, \xi, \vartheta) &= x_0(\bar{z}, \vartheta) + R(\bar{z}, \xi, \vartheta)\cos\vartheta \\
\bar{y}\,(\bar{z}, \xi, \vartheta) &= y_0(\bar{z}, \vartheta) + R(\bar{z}, \xi, \vartheta)\sin\vartheta \\
R\,(\bar{z}, \xi, \vartheta) &= G(\bar{z}, \vartheta) + \xi^n\,(F(\bar{z}, \vartheta) - G(\bar{z}, \vartheta))
\end{aligned}
\tag{4}
$$

Die Funktionen $x_0(\bar{z}, \vartheta)$ und $y_0\,(\bar{z}, \vartheta)$ sind frei wählbar und sollen dem jeweiligen Problem angepaßt werden. Die Berechnung der metrischen Ableitungen kann über die Funktionaldeterminante geschehen

$$
\left[\frac{\partial\,(z, r, \varphi)}{\partial\,(\bar{z}, \xi, \vartheta)}\right]^{-1} = \frac{\partial(\bar{z}, \xi, \vartheta)}{\partial(z, r, \varphi)}
$$

Die Ableitungen $\frac{\partial}{\partial z}, \frac{\partial}{\partial r}, \frac{\partial}{\partial \varphi}$ transformieren sich mit

$$
\begin{aligned}
\frac{\partial}{\partial z} &= \frac{\partial}{\partial \bar{z}} + \xi_z \frac{\partial}{\partial \xi} + \vartheta_z \frac{\partial}{\partial \vartheta} \\
\frac{\partial}{\partial r} &= \xi_r \frac{\partial}{\partial \xi} + \vartheta_r \frac{\partial}{\partial \vartheta} \\
\frac{\partial}{\partial \varphi} &= \xi_\varphi \frac{\partial}{\partial \xi} + \vartheta_\varphi \frac{\partial}{\partial \vartheta}
\end{aligned}
$$

Damit erhält man aus (3)

$$
J(U)\,\frac{\partial U}{\partial \bar{z}} + \overline{K(U)}\,\frac{\partial U}{\partial \xi} + \overline{L(U)}\,\frac{\partial U}{\partial \vartheta} + H(U) = 0 \tag{3a}
$$

$$
\overline{K(U)} = J(U)\xi_z + K(U)\xi_r + L(U)\xi_\varphi
$$

$$
\overline{L(U)} = J(U)\vartheta_z + K(U)\vartheta_r + L(U)\vartheta_\varphi
$$

IV. Numerisches Verfahren

Zur Beschreibung eines konkreten Problems - wie hier das der Umströmung eines Deltaflügels mit Oberschallanströmung - gehören noch Anfangs- und Randbedingungen. Da das Berechnungsgebiet durch die Körperkonturfläche und die Stoßkonturfläche begrenzt ist, müssen dort Randbedingungen angegeben werden, die am Körper durch seine Undurchlässigkeit gegeben ist

$\vec{v} \cdot \vec{n} = 0, \qquad \vec{n} \triangleq$ Körpernormale

und an der Stoßkonturfläche, die selbst unbekannt und Teil der Lösung ist, durch die Rankine-Hugoniot-Gleichungen [2].

Die Anfangsbedingungen werden am Orte $z = z_0$ durch eine Näherungslösung für das adäquate konische Problem gewählt.

Die numerische Approximation der Gleichung (3a) mit finiten Differenzen wird nach der Methode, welche in [5] und [2] beschrieben ist, durchgeführt. Es hat sich jedoch herausgestellt, daß Stabilität des Verfahrens beim Auftreten von eingebetteten Verdichtungsstößen nur dann zu erhalten ist, wenn der in [2] verwendete Differenzenoperator $\Delta_{x,j}$ durch

$$2\tau\left(\frac{\partial}{\partial \bar{z}}\right)^{n+(j/2)}_{m+1/2,l} \rightarrow \Delta_{\bar{z},j} = (P_1 + I)\left(P_0^j - I - \frac{\delta_1 \kappa_2}{4}(P_2 - 2I + P_2^{-1})\right) \qquad (5)$$

$$+ \frac{\kappa_1}{2h_1}(P_1 - I)\ \Theta_m(P_1 - P_1^{-1})$$

$$\Theta_m = \varepsilon \frac{(|v| + c)_m}{(|v| + c)_{max}}$$

ersetzt wird. Der zweite Summand in Gl. (5) bedeutet die Addition eines künstlichen Dissipationsterms in ξ-Richtung (gilt nur für Feldpunkte).

Aus den Randbedingungen am Stoß und dem Ergebnis der Rekursionsformel ist eine Gleichung für die Steigung der Stoßkontur herleitbar (siehe [2]). Diese ändert sich für den hier beschriebenen Fall gegenüber der Gleichung 4.4 aus [2] zu

$$\varphi(F_{\bar{z}}^{(s+1)}) = -\frac{1}{B}\left\{A + \frac{1}{m_{1;\infty}}\left[-\xi_r^{(s)} m_{2;\infty} - \frac{1}{r}\,\xi_\varphi^{(s)} m_{3;\infty} + \frac{(\beta^{(s)})^2}{\mu\,\nu}\,\psi^{(s)}\right]\right\} \qquad (6)$$

mit

$$A = \sin\vartheta\left[x_{o,\bar{z}} \cdot F_{\vartheta;l}^{n+(j)} + y_{o,\bar{z}}\, F_l^{n+(j)}\right]$$

$$+ \cos\vartheta\left[x_{o,\bar{z}} \cdot F_l^{n+(j)} - y_{o,\bar{z}}\, F_{\vartheta;l}^{n+(j)}\right]$$

$$+ x_{o,\bar{z}}\, y_{o,\vartheta} - y_{o,\bar{z}}\, x_{o,\vartheta}$$

$$B = F_l^{n+(j)} - \sin\vartheta\, x_{o,\vartheta} + \cos\vartheta\, y_{o,\vartheta}$$

$$\psi^{(s)} = \frac{g_M - g_\infty}{(1 - \rho^{(s)}/\rho_\infty)} + g_\infty - \mu_5\left\{e_\infty + \frac{(\alpha_\infty^{(s)})^2}{\rho_\infty(\beta^{(s)})^2}\left[\frac{1}{2}\left(\frac{\rho_\infty}{\rho^{(s)}} - 1\right) - \frac{\rho_\infty}{\rho^{(s)}(\gamma-1)}\right] - \frac{1}{2}\frac{m_\infty^2}{\rho_\infty}\right\}$$

Für weitere Einzelheiten sei auf [2] verwiesen.

V. Ergebnisse

Bild 3 zeigt das Ergebnis der Berechnung des Strömungsfeldes um einen Rotationskörper, der vorne mit einer Kugelnase abgestumpft ist. Für zwei verschiedene Machzahlen (M_∞ = 1.2; Bild 3a und M_∞ = 2.0; Bild 3b) sind die Stromlinien, die Schall-Linie und die Verdichtungsstoßkontur eingetragen. Man sieht, daß für kleiner werdende Machzahlen das Unterschallgebiet stark anwächst und der Verdichtungsstoß stromauf wandert. Für $M_\infty \to 1$ wird der Verdichtungsstoß in eine Mach'sche Linie übergehen und er wird im Unendlichen vor dem erzeugenden Körper stehen.

Bild 4 ist der Arbeit [3] entnommen und soll die Leistungsfähigkeit des verwendeten Verfahrens verdeutlichen. Die Anströmmachzahl beträgt $M_\infty = 1.05$. Bild 4a behandelt den Fall der Halbkugel-Zylinderumströmung, während in Bild 4b die Kugelumströmung dargestellt ist. Aus physikalischen Gründen müssen die Staupunktsbereiche der beiden Strömungsfelder dann identisch sein, wenn die vom Körper ausgehende Grenzcharakteristik in beiden Fällen vom vorderen Kugelteil ausgeht. Da jedoch die Position und die Form der Grenzcharakteristik Teil der Feldlösung sind, kann diese Frage erst nach der erfolgreichen Berechnung der beiden Strömungsfelder beantwortet werden. Zur Berechnung der Halbkugel-Zylinderströmung wurde das Koordinatensystem aus [1] verwendet, während für die Kugelströmung das Koordinatensystem aus [3] Anwendung fand. Die Rechnungen verliefen in beiden Fällen stabil und konvergierten zur zeitlich asymptotischen Lösung ([3 , 8]). Die Auswertung der Strömungsfelder zeigte nun, daß die Grenzcharakteristik in der Tat in beiden Fällen vom vorderen Teil der Kugel ausgeht, sodaß das Strömungsfeld im Staupunktsbereich beider Körper identisch sein muß. Dies wird durch die Bilder 4a und 4b, in denen Kopfwelle, Schall-Linie und Charakteristiken eingetragen sind, sehr gut bestätigt.

Die nächste Abbildung (Bild 5) zeigt das Strömungsfeld im Bereich reiner Überschallströmung um einen aus zylindrischen und konischen Teilstücken zusammengesetzten Körper, der vorne mit einer Kugelnase abgestumpft ist. Der Staupunktbereich wurde mit dem 3-D "time-marching"-Verfahren berechnet. Mit dieser Lösung sind gleichzeitig auch die Anfangsdaten für das 3-D "space-marching"-Verfahren gegeben. Bei der Umströmung einer Geometrie der Art, wie sie in Bild 5 gezeigt ist, entstehen aufgrund von Rekompressionen bzw. von konkaven Ecken eingebettete Verdichtungsstöße. Diese lassen sich in 2-D Strömungen bzw. in Symmetrieebenen von 3-D Strömungen durch den Verlauf der Charakteristiken, die hier mit den Mach'schen Linien identisch sind, sichtbar machen. In Bereichen konvergierender Charakteristiken wird das Strömungsmedium komprimiert, während bei divergierenden Charakteristiken Expansion auftritt. Das Bild 5 zeigt in der Symmetrieebene der Strömung ($M_\infty = 2.0$; α = 10°) die Kopfwelle und die Stromlinien 5a sowie die Charakteristiken 5b. Man kann sehr schön die aufeinanderfolgenden Bereiche der Expansion und der Kompression der Strömung sowohl luv- als auch leeseitig sehen.

Das Bild 6 zeigt ein Koordinatennetz, wie es mit den Transformationsgleichungen (4) erzeugt worden ist. Die Funktionen x_0 und y_0 wurden dabei durch

$$x_0\,(\bar{z},\vartheta) = 0$$

$$y_0(\bar{z},\vartheta) = \tan\delta_2 \cdot \sin\vartheta \cdot \bar{z}$$

$$\tan\delta_2 = 0.450$$

vorgegeben. Die Vorzüge eines solchen Gitternetzes sind offensichtlich.

Die Umströmung des konischen Teils eines Deltaflügels wird in den Bildern 7 und 8 dargestellt. Es zeigt Bild 7, neben der Kopfwelle, Linien konstanten statischen Drucks (Isobaren) für den Anströmfall $M_\infty = 2.0$ und Anstellwinkel $\alpha = 10°$. An der abgestumpften Flügelvorderkante ergibt sich eine erhebliche Druckabsenkung, die sich auf der Leeseite fortsetzt. Die Vektoren der Querströmung sind in Bild 8 aufgetragen. Das Strömungsfeld um einen elliptischen Konus mit dem Achsenverhältnis 1:8 und der Gleichung $\frac{x^2}{a^2} + \frac{y^2}{b^2} = \frac{z^2}{c^2}$ ($a = 0.25$; $b = 2$; $c = 4$) ist bei den Anströmverhältnissen $M_\infty = 2.0$ und $\alpha = 0°$ berechnet worden. Das Bild 9 zeigt dazu die Linien konstanter Gesamtenergie, die durch

$$E = e_i + \frac{1}{2}\rho\,(u^2 + v^2 + w^2) \qquad e_i \text{ innere Energie}$$

definiert ist. Man beachte, daß für eine stationäre Strömung nicht E sondern die Gesamtenthalpie H die Erhaltungsgröße ist.

Bild 10 zeigt einen Deltaflügel, der bis z = 1.0 einen konischen Verlauf hat und anschließend eine konstante Dicke besitzt, wobei die stumpfe Flügelkante durch ein 1:2 Ellipsoid gebildet wird. Die letzte Abbildung (Bild 11) zeigt die Vektoren der Querströmung bei z = 2.0. Die Verdrängung der Strömung, wie sie im konischen Teil stattfindet (Bild 8) wird hier nicht beobachtet. Vielmehr wird von der Luvseite kommend die Strömung um die Flügelvorderkante gelenkt und strömt mit abnehmender Stärke längs der Konturfläche in Richtung Symmetrieebene.

Die reibungsbehaftete Strömung wird sich anders verhalten; sie wird im Bereich der Vorderkante aufgrund des zu beobachtenden Druckanstieges ablösen und es werden auf der Leeseite Wirbelstrukturen entstehen. Für scharfe Vorderkanten ist dieses Phänomen bei Verwendung einer Kutta-Bedingung an der Vorderkante auch im Rahmen einer reibungsfreien Lösung (Integration der Euler-Gleichungen) zu beobachten [9].

Literatur:

[1] Weiland C.: ZfW 24 (1976), pp. 237 ÷ 245

[2] Weiland C.: J. Comp. Phys. 29 (1978), pp. 173 ÷ 198

[3] Weiland C.: Comp. & Fluids 9 (1981), pp. 143 ÷ 162.

[4] Lyubimow A.N.; Rusanow V.V.: NASA TT F-714 (1973).

[5] Babenko K.I.; Voskresenskii G.P.; Lyubimow A.N.; Rusanow V.V.: NASA TT F-380 (1966).

[6] Lax P.D.: Com. on Pure & Appl. Mathe. 7 (1954), pp. 159 ÷ 193.

[7] Weiland C.: Über eine der klassischen Erhaltungssatzform äquivalenten Formulierung der Euler'schen Bewegungsgleichungen zur Berechnung eingebetteter Verdichtungsstöße (wird veröffentlicht).

[8] Weiland C.: In: Recent Developments in Theoretical and Experimental Fluid Mechanics (Müller U.; Roesner K.G.; Schmidt B.; Eds.), Berlin: Springer-Verlag (1979).

[9] Weiland C.: Vortex-Flow Simulation Past Wings Using the Euler Equations (wird veröffentlicht).

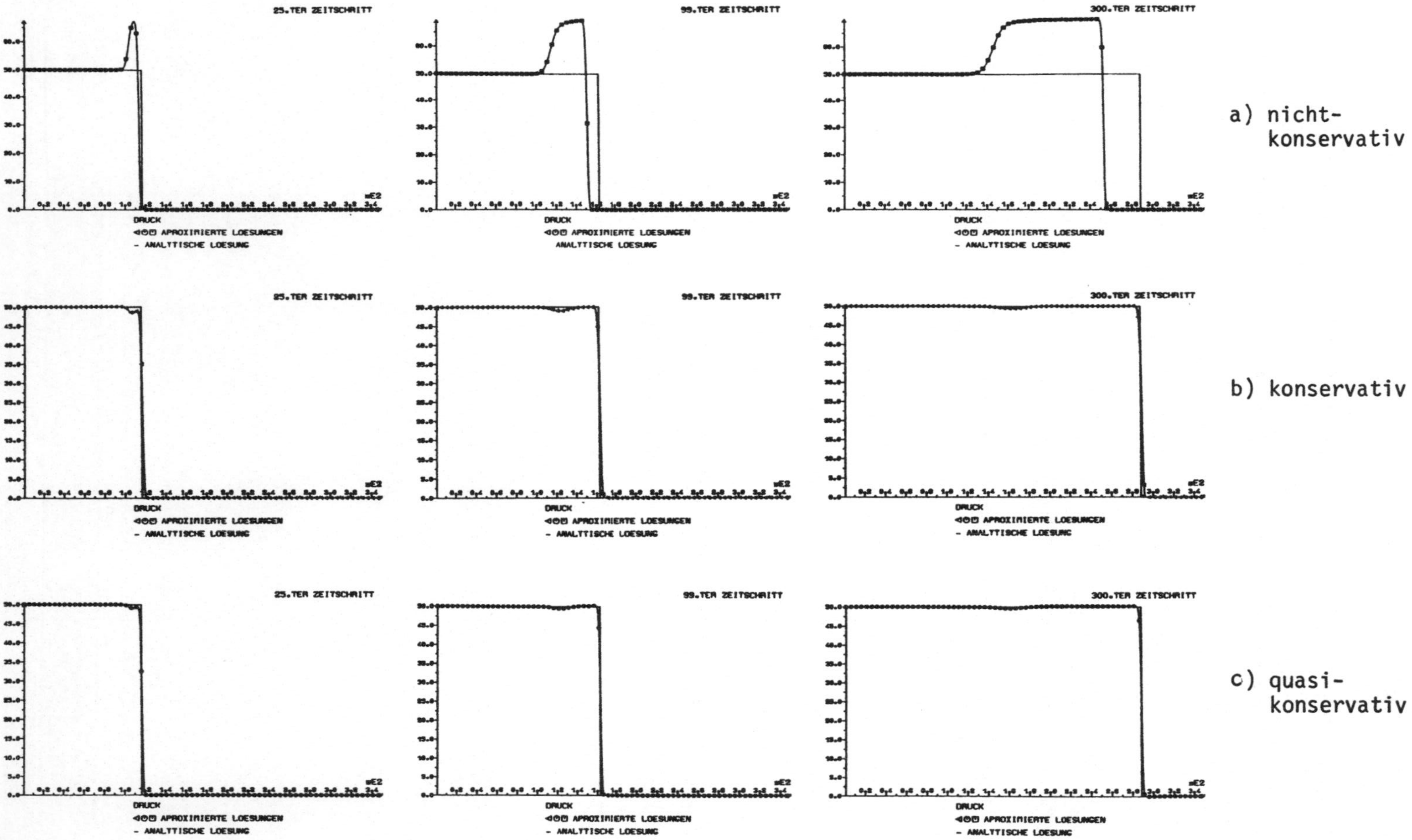
25.TER ZEITSCHRITT
99.TER ZEITSCHRITT
300.TER ZEITSCHRITT
DRUCK
APROXIMIERTE LOESUNGEN
- ANALYTISCHE LOESUNG
*E2
a) nicht-
konservativ
b) konservativ
c) quasi-
konservativ

B i l d 1

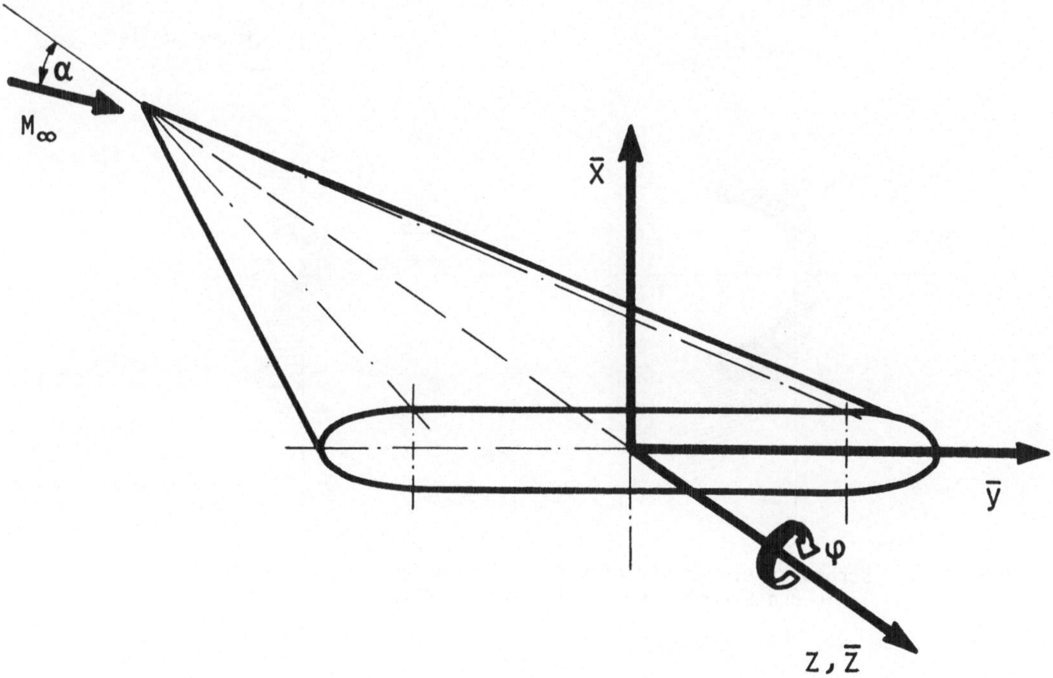

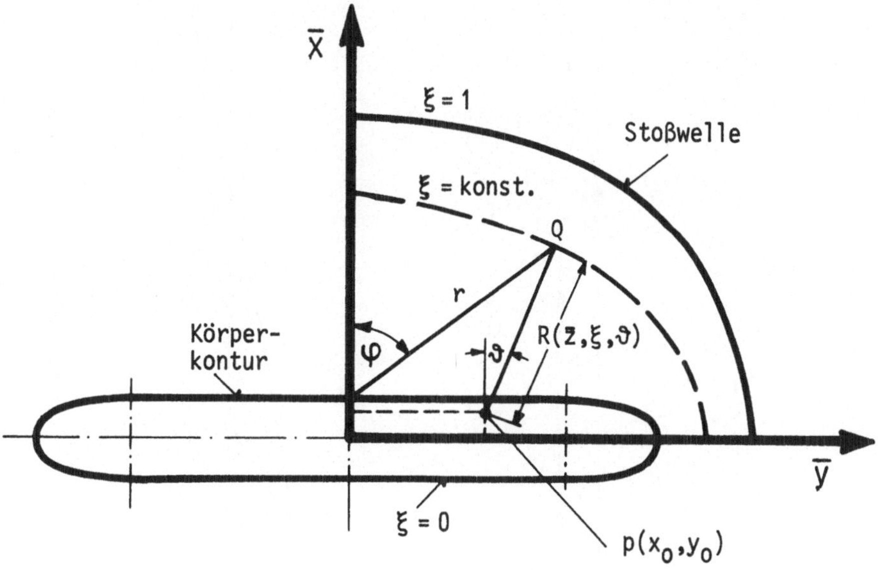

$$R\ (\bar{z},\xi,\vartheta) = G\ (\bar{z},\vartheta) + \xi^{n}\ (F\ (\bar{z},\vartheta) - G\ (\bar{z},\vartheta))$$

$$x_0 = x_0\ (\bar{z},\vartheta)$$

$$y_0 = y_0\ (\bar{z},\vartheta)$$

Bild 2

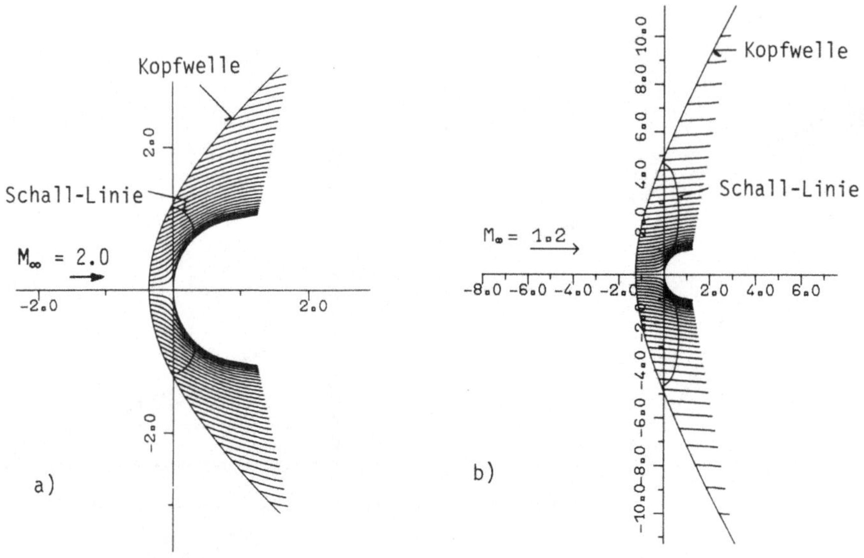

Bild 3 Strömungsfeld um stumpfen Rotationskörper, Kopfwelle, Schall-linie und Stromlinien im Staupunktsbereich

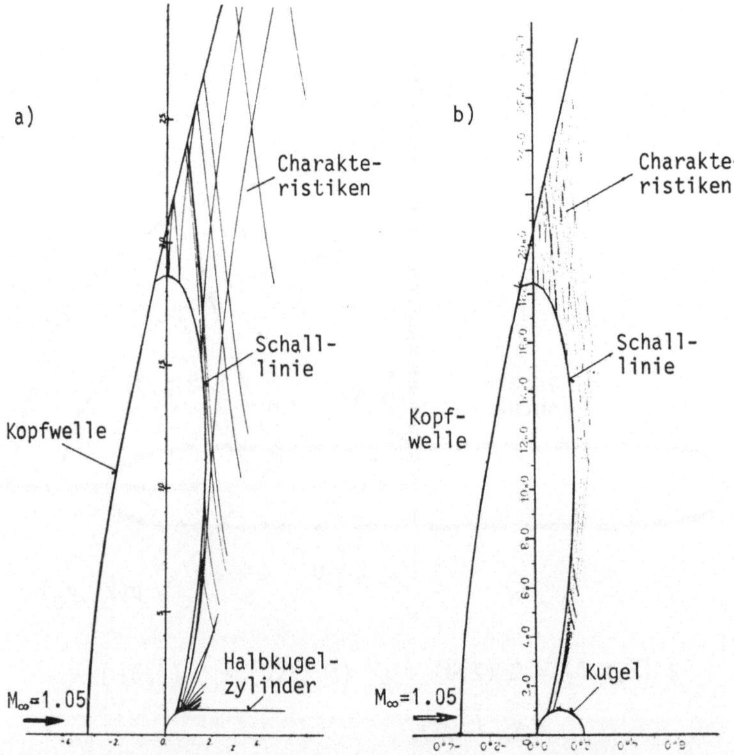

Bild 4 Strömungsfeld um a) Halbkugelzylinder, b) Kugel bei M_∞ = 1.05, Charakteristiken mit Grenzcharakteristik im Überschall

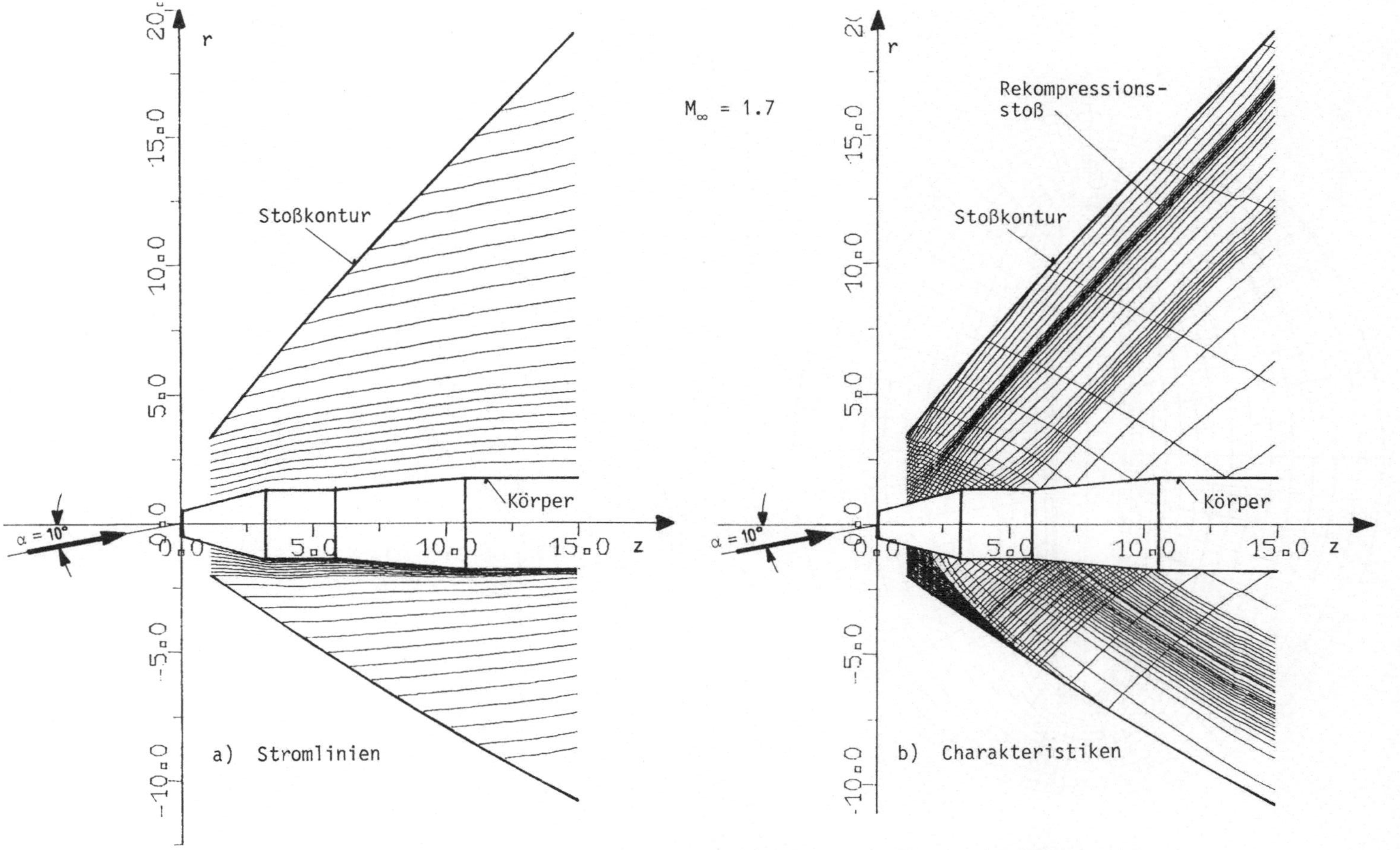

Bild 5 Darstellung von Stromlinien und Charakteristiken in der Symmetrieebene des Strömungsfeldes

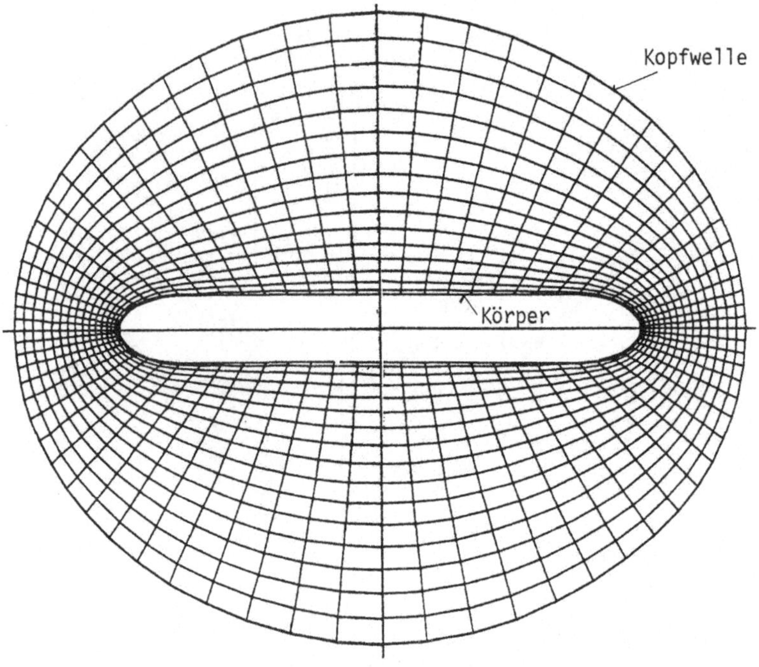

Bild 6 Koordinatennetz in Ebenen z = konstant eines Delta-Flügels

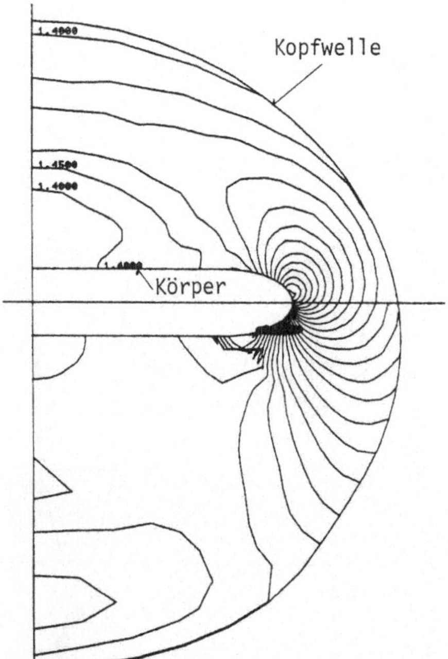

Bild 7 Linien konstanten statischen Drucks in Ebenen z = konstant für konischen Delta-Flügel mit $M_\infty = 2.0$, Anstellwinkel $\alpha = 10°$

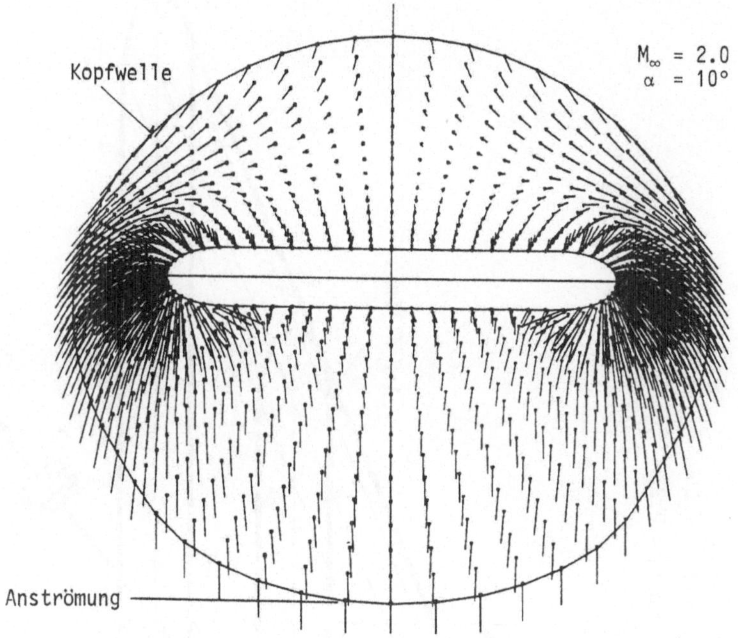

Bild 8 Vektoren der Querströmung in Ebenen z = konstant des konischen Teils eines Delta-Flügels

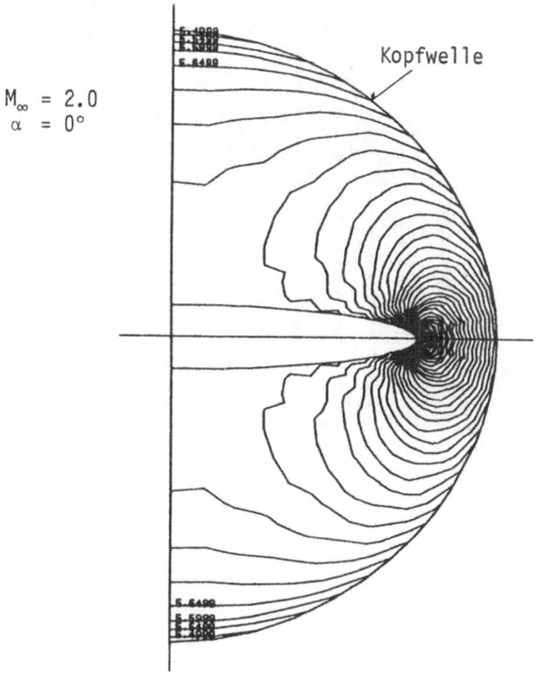

Bild 9 Elliptischer Konus mit einem Achsenverhältnis von 1:8; Linien konstanter spezifischer Gesamtenergie E

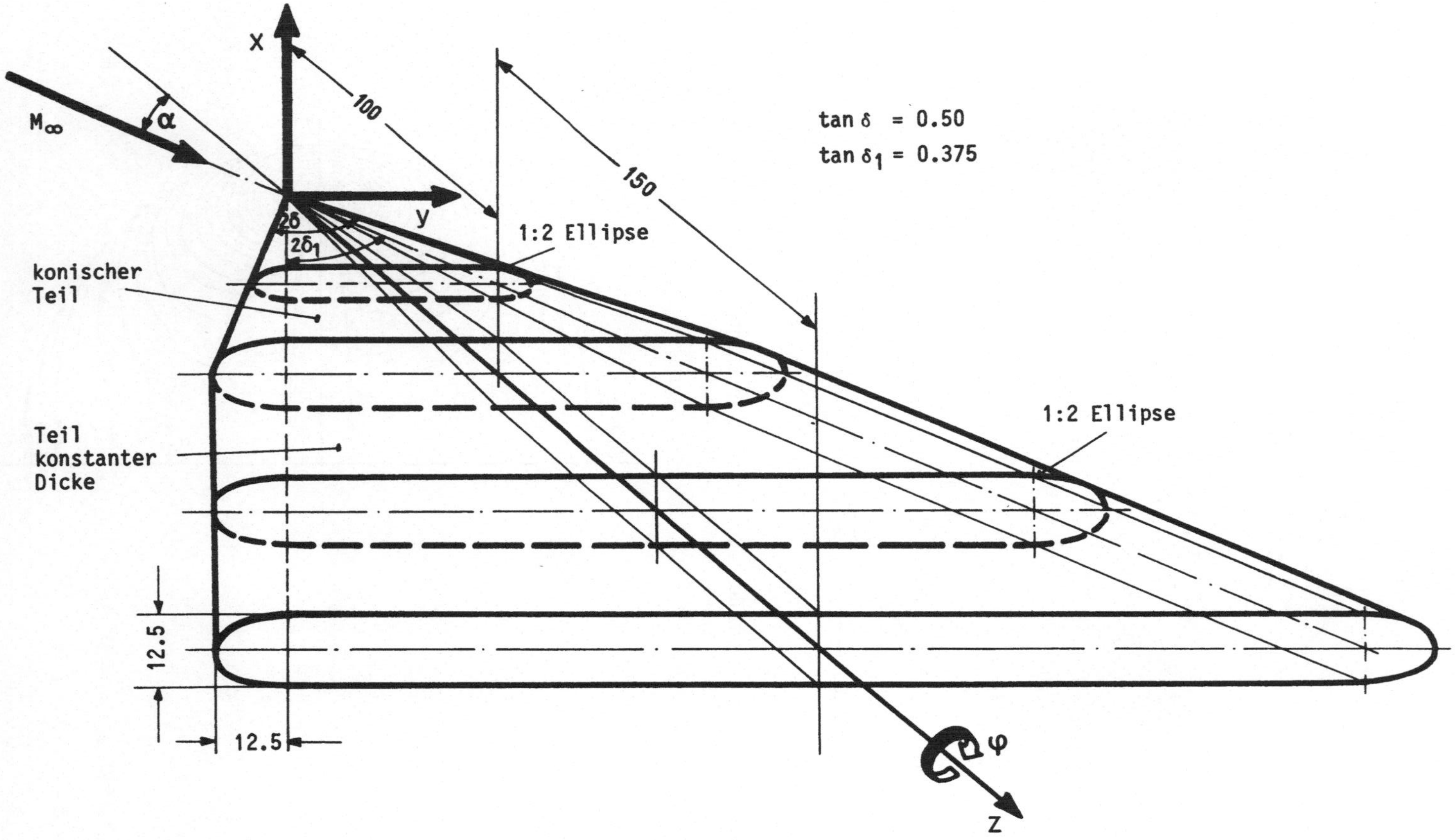
x
y
z
φ
M_∞
α
100
150
tan δ = 0.50
tan δ₁ = 0.375
2δ
2δ₁
1:2 Ellipse
1:2 Ellipse
konischer Teil
Teil konstanter Dicke
12.5
12.5

Bild 10

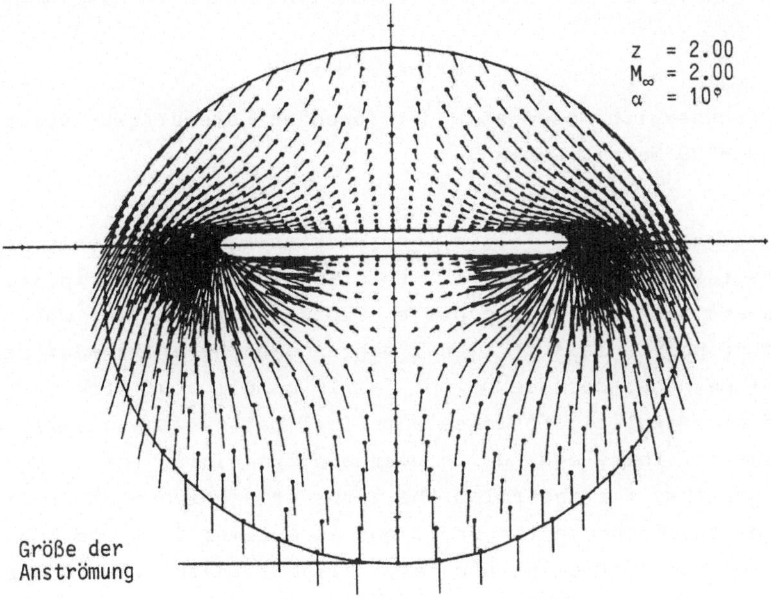

Bild 11 Vektoren der Querströmung in der Ebene z = 2.0, Delta-Flügel mit stumpfer Vorderkante

ON THE SIMULTANEOUS DETERMINATION OF POLYNOMIAL ROOTS

Wilhelm Werner

Fachbereich Mathematik der Johannes Gutenberg-Universität
D 6500 Mainz, Germany

Introduction. Galois' famous theorem states that the only polynomial equations that possess a general solution in terms of "explicit" formulas are those of degree not exceeding four. For the computation of zeros of polynomials of higher degree one therefore *must* resort to numerical methods; because of the fairly complicated structure of the corresponding explicit formulas numerical methods are usually employed in the case of third and fourth degree polynomials, too.

Many algorithms for determing the zeros of a polynomial are known. Most algorithms compute the zeros one at a time; if a root has been determined with sufficient accuracy a linear factor is removed from the polynomial by the Horner algorithm and the algorithm is used again to compute a zero of the deflated polynomial. The method of successive removal of linear factors has the following disadvantage: In applications sometimes it is not necessary to determine the zeros to any great accuracy. If the method of successive deflation is used, evidently it is not possible to take advantage of this special situation for if the early linear factors are very inaccurate the polynomials obtained by successive deflation may be falsified to an extend that makes the remaining approximate zeros meaningless.

An alternative approach which overcomes the difficulties mentioned above uses the fact that by the *Fundamental Theorem of Algebra* the number of solutions of a polynomial equation is known in advance. One therefore associates with a given polynomial p a *system of nonlinear equations* whose solution is a vector consisting of the roots of p. This device is considered in the present note; it is shown how the computational efficiency of various known methods for the simultaneous determination of polynomial roots can be substantially improved. These improvements are achieved partly by a computational more efficient realization of known methods, partly by a modification of existing methods such that their order of convergence is increased without relevant additional computations.

1. *Some remarks on polynomial interpolation*

In the sequel we always assume that p is a (complex) polynomial given in the form

$$(1)\qquad p(z) = \sum_{i=0}^{n} a_i \prod_{j=0}^{i-1} (z-z_j) \quad , \; a_i \in \mathbb{C},\; a_n \neq 0,\; z_i \in \mathbb{C},$$

i.e. $p \in \Pi_n$, where Π_n denotes the space of polynomials of degree $\leq n$. If p is to be represented in the form

$$(2)\qquad p(z) = \sum_{i=0}^{n} \alpha_i \prod_{j=0}^{i-1} (z-\zeta_j)$$

for some given $\zeta_o, \zeta_1, \ldots, \zeta_{n-1} \in \mathbb{C}$, then the unknown coefficients α_i can be computed economically by the

general Horner algorithm:

$$a_k^{(o)} := a_k, \qquad k=0(1)n$$

$$(3)\qquad \left.\begin{aligned} a_n^{(i+1)} &:= a_n^{(i)} \\ a_k^{(i+1)} &:= a_k^{(i)} + (\zeta_i - z_{k-i})a_{k+1}^{(i+1)} \quad , \; k=n-1(-1)i \\ \alpha_i &:= a_i^{(i+1)} \end{aligned}\right\} \; i=0(1)n.$$

PROPOSITION 1. *If* $q_i \in \Pi_{n-i}$, $i=0(1)n$, *is defined by*

$$q_i(z) := \sum_{k=i}^{n} a_k^{(i)} \prod_{j=0}^{k-i-1} (z-z_j), \; z \in \mathbb{C},$$

then $\quad q_i(z) = a_i^{(i+1)} + (z-\zeta_i)q_{i+1}(z), \quad i=n-1(-1)0.$

PROOF.

$$(z-\zeta_i)q_{i+1}(z) = (z-\zeta_i) \sum_{k=i+1}^{n} a_k^{(i+1)} \prod_{j=0}^{k-i-2} (z-z_j)$$

$$= \sum_{k=i}^{n-1} a_{k+1}^{(i+1)} \{(z-z_{k-i})+(z_{k-i}-\zeta_i)\} \prod_{j=0}^{k-i-1} (z-z_j)$$

$$= \sum_{k=i}^{n-1} a_{k+1}^{(i+1)} \prod_{j=0}^{k-i} (z-z_j) + \sum_{k=i}^{n-1} a_{k+1}^{(i+1)} (z_{k-i}-\zeta_i) \prod_{j=0}^{k-i-1} (z-z_j)$$

$$= \sum_{k=i+1}^{n} a_k^{(i+1)} \prod_{j=0}^{k-i-1} (z-z_j) + \sum_{k=i}^{n-1} (a_k^{(i)}-a_k^{(i+1)}) \prod_{j=0}^{k-i-1} (z-z_j)$$

$$= q_i(z) - a_i^{(i+1)}$$

COROLLARY 2. $q_i \in \Pi_{n-i}$ *is the i-th divided difference of* p *with respect to* $\zeta_o, \zeta_1, \ldots, \zeta_{i-1}$:

$$q_o(z) = p(z)$$

$$q_i(z) = p[\zeta_o, \ldots, \zeta_{i-1}, z] \qquad (i=1(1)n);$$

especially it is

$$p(z) = \sum_{i=0}^{n} a_i^{(i+1)} \prod_{j=0}^{i-1} (z-\zeta_j) .$$

In the sequel let us assume that $z_i \neq z_j$ for $i \neq j$ in (1); if p is to be transformed from the Newtonian form (1) into the Lagrangian form of a polynomial (with remainder)

$$(4) \qquad p(z) = \sum_{i=0}^{n-1} \frac{p(z_i)}{\prod_{\substack{j=0\\ j\neq i}}^{n-1} (z_i-z_j)} \prod_{\substack{j=0\\ j\neq i}}^{n-1} (z-z_j) + a_n \prod_{j=0}^{n-1} (z-z_j)$$

then the coefficients

$$(5) \qquad \sigma_i := p(z_i) / \prod_{\substack{j=0\\ j\neq i}}^{n-1} (z_i-z_j) \qquad (i=0(1)n-1)$$

may be computed efficiently by the following algorithm:

$$a_k^{(o)} := a_k \qquad (k=0(1)n)$$

$$(6) \qquad \left.\begin{aligned} a_k^{(i)} &:= a_k^{(i-1)}/(z_k-z_i) \\ a_i^{(k+1)} &:= a_i^{(k)} - a_k^{(i)} \end{aligned}\right\} \quad i=k+1(1)n-1,\ k=0(1)n-2$$

$$\sigma_k := a_k^{(n-1)} \qquad k=0(1)n-1$$

This is proved in the following

PROPOSITION 3. *If* $z_i \neq z_j$ *for* $i \neq j$ *then*

$$\sum_{i=0}^{n-1} a_i \prod_{j=0}^{i-1} (z-z_j) = \sum_{i=0}^{n-1} a_i^{(n-1)} \prod_{\substack{j=0\\ j\neq i}}^{n-1} (z-z_j)$$

PROOF. The following assertion will be proved by induction:

$$(7)\qquad \left.\begin{aligned} a_k^{(l)} &= \sum_{i=l}^{k} \sigma_i \prod_{j=k+1}^{n-1} (z_i-z_j) \qquad && (k=l(1)n-1) \\ a_k^{(n-1)} &= \sigma_k && (k=0(1)l-1) \end{aligned}\right\} \quad l=0(1)n \quad .$$

By a well known result on divided differences (cf. MILNE-THOMPSON [21], p. 7) one has

$$a_k^{(o)} = \sum_{i=0}^{k} p(z_i)/ \prod_{\substack{j=0\\ j\neq i}}^{k} (z_i-z_j)$$

$$(8)\qquad = \sum_{i=0}^{k} \sigma_i \prod_{j=k+1}^{n-1} (z_i-z_j) \qquad \text{(by (5))},$$

so that (7) is valid for $l=0$.

By (6) $a_o^{(i)} = a_o^{(o)}/ \prod_{j=1}^{i} (z_o-z_j)$ $(i=0(1)n-1)$ so that

$$a_o^{(n-1)} = a_o/ \prod_{j=1}^{n-1} (z_o-z_j) = \sigma_o;$$

$$\begin{aligned} a_k^{(1)} &= a_k^{(o)} - a_o^{(k)} \\ &= a_k^{(o)} - a_o^{(o)}/ \prod_{j=1}^{k} (z_o-z_j) \\ &= a_k^{(o)} - \sigma_o \prod_{j=k+1}^{n-1} (z_o-z_j) \\ &= \sum_{i=1}^{k} \sigma_i \prod_{j=k+1}^{n-1} (z_i-z_j) \qquad \text{(by (8))}. \end{aligned}$$

Hence (7) is valid for $l=1$; a repetition of this argument yields the assertion.

REMARK. Algorithm (6) requires about $\frac{1}{2}n^2$ divisions and n^2 additions/subtractions; a straightforward computation of the σ_i $(i=0(1)n-1)$ from the definition (5) requires about $\frac{3}{2}n^2$ multiplications and $2n^2$ additions/subtractions.

2. *The DURAND-KERNER-method*

For convenience we assume in the following that $a_n = 1$ in (1); we consider the mapping

$$F_p : \mathbb{C}^n \to \Pi_{n-1},$$

$$F_p(z_0,z_1,\ldots,z_{n-1}) := p - \omega(z_0,z_1,\ldots,z_{n-1}) \tag{9}$$

where $\omega(z_0,\ldots,z_{n-1}) \in \Pi_n$ is defined by

$$\omega(z_0,\ldots,z_{n-1})(z) = \prod_{j=0}^{n-1} (z-z_j), \quad z \in \mathbb{C}.$$

Obviously the following statements are equivalent:

(10) i) $F_p(z_0,\ldots,z_{n-1}) = 0$

ii) $z_0, z_1, \ldots, z_{n-1}$ are the n roots of p .

This suggests to apply Newton's method to the nonlinear equation (10) in order to determine simultaneously the roots of p which we will denote by $\xi_0, \xi_1, \ldots, \xi_{n-1}$.

LEMMA 4. *If* Π_{n-1} *is considered as a finite-dimensional normed vectorspace then the Fréchet-derivative of* F_p *at* $(z_0,\ldots,z_{n-1})$ *is given by*

$$[F_p'(z_0,\ldots,z_{n-1})s](z) = \sum_{i=0}^{n-1} s_i \prod_{\substack{j=0\\ j\neq i}}^{n-1} (z-z_j),$$

for any $s = (s_0,\ldots,s_{n-1})^T \in \mathbb{C}^n$, $z \in \mathbb{C}$.

The PROOF results immediately from the definitions.

Hence we get

COROLLARY 5. *If* $z_i \neq z_j$ *for* $i \neq j$ *then the Newton-correction* $\sigma \in \mathbb{C}^n$,

$$F_p'(z_0,\ldots,z_{n-1})\sigma = F_p(z_0,\ldots,z_{n-1}),$$

can be uniquely determined from the equation

$$\sum_{i=0}^{n-1} \sigma_i \prod_{\substack{j=0\\ j\neq i}}^{n-1} (z-z_j) = \sum_{i=0}^{n-1} a_i \prod_{j=0}^{i-1} (z-z_j), \qquad z \in \mathbb{C}. \tag{11}$$

If $z_0^{(k)},\dots,z_{n-1}^{(k)} \in \mathbb{C}$ are pairwise distinct approximations for the roots $\xi_0,\dots,\xi_{n-1}$ of p, then (4), (5) and (11) imply that

$$(12) \qquad z_i^{(k+1)} := z_i^{(k)} - \frac{p(z_i^{(k)})}{\prod\limits_{\substack{j=0\\ j\neq i}}^{n-1} (z_i^{(k)}-z_j^{(k)})} \qquad (i=0(1)n-1)$$

actually describes a Newton-step for F_p.

REMARK.

(a) (12) was independently proposed by DURAND [10], p.279, KERNER [18] and DOCHEV-ILLIEF [9]. DURAND and KERNER derived (12) by an application of Newton's method to the elementary symmetric functions of $z_0^{(k)},\dots,$ $z_{n-1}^{(k)}$. Formerly (12) was used by WEIERSTRASS [25] for a proof of the fundamental theorem of algebra.

(b) A straightforward calculation of the Newton-corrections

$$(13) \qquad \sigma_i^{(k)} := p(z_i^{(k)})/\prod\limits_{\substack{j=0\\ j\neq i}}^{n-1} (z_i^{(k)}-z_j^{(k)}) \qquad (i=0(1)n-1)$$

requires about $2n^2$ multiplications and $3n^2$ additions/subtractions if p (as given by (1)) is evaluated by Horner's method.

A computational more economic realization of the DURAND-KERNER-method proceeds as follows:

1. Compute the coefficients $\alpha_i^{(k)}$ of the representation

$$p(z) = \sum_{i=0}^{n-1} \alpha_i^{(k)} \prod_{j=0}^{i-1} (z-z_j^{(k)}) + \prod_{j=0}^{n-1} (z-z_j^{(k)})$$

by the general Horner algorithm (3).

2. Compute the Newton corrections $\sigma_i^{(k)}$ according to (6) .

One step of the DURAND-KERNER-method thus can be performed with (about) $\frac{1}{2}n^2$ multiplications, $\frac{1}{2}n^2$ divisions and $2n^2$ additions/subtractions.

(c) A single step version of (12) is mentioned in ALEFELD-HERZBERGER [3]; in connection with interval analysis (12) is treated in ALEFELD-HERZBERGER [2].

Let us mention a useful property of the iterative method (12) (see also DOCHEV-ILLIEF [9], BYRNEV-DOCHEV [8]):

LEMMA 6. *Let* $\xi_0,\dots,\xi_{n-1} \in \mathbb{C}$ *be the roots of* p; *if* $z_i^{(k)} \neq z_j^{(k)}$ *for* $i \neq j$ *then*

$$\sum_{i=0}^{n-1} z_i^{(k+1)} = \sum_{i=0}^{n-1} \xi_i \;, \quad k=0,1,2,\dots \quad .$$

PROOF. For fixed m, $0 \le m \le n-1$, (11) implies that

$$\sum_{i=0}^{n-1} \sigma_i^{(k)} \prod_{\substack{j=0\\ j\neq i}}^{n-1} (\xi_m - z_j^{(k)}) = - \prod_{j=0}^{n-1} (\xi_m - z_j^{(k)}) \;;$$

if $v(\xi_m) := \Big(\prod_{\substack{j=0\\ j\neq i}}^{n-1} (\xi_m - z_j^{(k)}) \Big)^T_{i=0(1)n-1} \in \mathbb{C}^n$ then this is equivalent to the statement that

$$(14)\quad \left\{ \begin{array}{l} \xi_m \text{ is an eigenvalue of} \\ A_p := \operatorname{diag}(z_0^{(k)},\dots,z_{n-1}^{(k)}) - (1,\dots,1)^T(\sigma_0^{(k)},\dots,\sigma_{n-1}^{(k)}) \in \mathbb{C}^{n,n} \\ \text{belonging to the eigenvector } v(\xi_m). \end{array} \right.$$

Hence trace $A_p = \sum_{i=0}^{n-1} (z_i^{(k)} - \sigma_i^{(k)}) = \sum_{i=0}^{n-1} \xi_i$.

REMARK. (14) was used by SMITH [24] and ELSNER [12] for the proof of error estimates via Gershgorin's disc theorem:

all roots of p are contained in the union of the discs
$D_j := \{ z \in \mathbb{C} \mid |z - z_j^{(k)} + \sigma_j^{(k)}| \le (n-1)|\sigma_j^{(k)}| \}$, $j=0(1)n-1$

(see also BRAESS-HADELER [6]). This fact may be used e.g. as a stopping criterion for the DURAND-KERNER-method.

There are various other possible derivations of the DURAND-KERNER-method which are not based on Newton's method; an example is discussed below.

As is obvious from (4), (5)

$$(15)\qquad \sigma_i := \sigma_i(z_0,\dots,z_{n-1}) = 0 \qquad (i=0(1)n-1)$$

implies that $z_0,\dots,z_{n-1} \in \mathbb{C}$ are the roots of p if p has n distinct zeros. It is therefore nearby to apply Newton-like methods to the nonlinear system (15). For this purpose it is necessary to compute the corresponding Jacobian:

LEMMA 7. *Let* $z_i \neq z_j$ *for* $i \neq j$; *then*

(a)
$$\frac{\partial\sigma_i}{\partial z_l} = \begin{cases} \dfrac{\sigma_i}{z_i - z_j} & \text{if } i \neq l \\ 1 - \sum\limits_{\substack{j=0\\ j\neq l}}^{n-1} \dfrac{\sigma_j}{z_j - z_l} & \text{if } i = l \end{cases}$$

(b)
$$\frac{\partial\sigma_l}{\partial z_l} = \sigma_l\left(\frac{p'(z_l)}{p(z_l)} - \sum_{\substack{j=0\\ j\neq l}}^{n-1} \frac{1}{z_l - z_j} \right) .$$

PROOF. From

$$p(z) = \sum_{i=0}^{n-1} \sigma_i \prod_{\substack{j=0\\ j\neq i}}^{n-1} (z-z_j) + \prod_{j=0}^{n-1} (z-z_j)$$

one concludes by differentiation that

$$0 = \sum_{\substack{i=0\\ i\neq l}}^{n-1} \Big[\frac{\partial\sigma_i}{\partial z_l} \prod_{\substack{j=0\\ j\neq i}}^{n-1} (z-z_j) - \sigma_i \prod_{\substack{j=0\\ j\neq i,l}}^{n-1} (z-z_j)\Big] + \Big(\frac{\partial\sigma_l}{\partial z_l} - 1 \Big) \prod_{\substack{j=0\\ j\neq l}}^{n-1} (z-z_j) ;$$

inserting z_k, $k=0(1)n-1$, yields (a). Differentiation of the equation

$$\frac{p(z)}{\prod\limits_{\substack{j=0\\ j\neq l}}^{n-1} (z-z_j)} = (z-z_l)\Big[1 + \sum_{\substack{i=0\\ i\neq l}}^{n-1} \frac{\sigma_i}{z-z_i}\Big] + \sigma_l \tag{16}$$

with respect to z and inserting z_l yields (b).

If $\tilde{x}$ is a zero of the twice differentiable mapping $F: X \to X$, X any suitable Banach space, then the iterative method

$$x_{n+1} := x_n - F'(\tilde{x})^{-1} F(x_n) \qquad (n=0,1,2,3,\ldots) \tag{17}$$

locally converges quadratically to $\tilde{x}$. By Lemma 7 one may interpret the DURAND-KERNER-method as an application of the Newton-like method (17) to the nonlinear system of equations

$$\sigma_i(z_0,\ldots z_{n-1}) = 0 \qquad (i=0(1)n-1) ,$$

too.

3. Higher order methods

Formula (16) is a well suited starting point for the construction of higher order methods: if z is a zero of p then

$$(18)\qquad z = z_l - \frac{\sigma_l}{1+\sum\limits_{\substack{i=0\\ i\neq l}}^{n-1}\frac{\sigma_i}{z-z_i}} .$$

This fixed point equation suggests the third order method

$$(19)\qquad z_l^{(k+1)} = z_l^{(k)} - \sigma_l^{(k)}/\Big(1+\sum_{\substack{i=0\\ i\neq l}}^{n-1}\frac{\sigma_i^{(k)}}{z_l^{(k)}-z_i^{(k)}}\Big)$$

$$(19^*)\qquad = z_l^{(k)} - \frac{p(z_l^{(k)})}{p'(z_l^{(k)})}/\Big(1-\frac{p(z_l^{(k)})}{p'(z_l^{(k)})}\sum_{\substack{j=0\\ j\neq l}}^{n-1}\frac{1}{z_l^{(k)}-z_j^{(k)}}\Big)$$

$l=0(1)n-1,\ k=0,1,2,3,\ldots$

(The equivalence of (19) and (19^*) follows from lemma 7.)

REMARK.
BÖRSCH-SUPAN [5] proposed to use (19) for the simultaneous determination of polynomial roots; (19^*) was used in different contexts: BÖRSCH-SUPAN [4] (a posteriori error bounds), ABERTH [1] , EHRLICH [11], GARGANTINI-HENRICI [13], HENRICI [15] (application of circular arithmetic). A single step version of (19^*) was investigated in ALEFELD-HERZBERGER [3].

Note however that (19^*) requires about $2n^2$ multiplications and n^2 divisions per step (if p and p' are evaluated by the Horner algorithm) whereas one step of (19) can be performed with (about) $\frac{1}{2}n^2$ multiplications and $\frac{3}{2}n^2$ divisions only if algorithm (6) is used for the computation of $\sigma_i^{(k)}$, $i=0(1)n-1$.

A slight modification of (19) improves the order of convergence:

$$(20)\qquad z_l^{(k+1)} = z_l^{(k)} - \sigma_l^{(k)}/\Big(1+\sum_{\substack{i=0\\ i\neq l}}^{n-1}\frac{\sigma_i^{(k)}}{z_l^{(k)}-\sigma_l^{(k)}-z_i^{(k)}}\Big)$$

$l=0(1)n-1,\ k=0,1,2,3,\ldots$

as is proved in the following proposition.

PROPOSITION 9. *If* $p \in \Pi_n$ *has n distinct roots* $\xi_0, \dots, \xi_{n-1}$, *then the iterative method* (20) *locally is of R-order 4* .

PROOF. For a suitable numeration of the roots the following equations are valid:

$$z_1^{(k+1)}-\xi_1 = z_1^{(k)}-\xi_1 - \sigma_1^{(k)}/\left(1+\sum_{i\neq 1}\frac{\sigma_i^{(k)}}{z_1^{(k)}-\sigma_1^{(k)}-z_i^{(k)}}\right)$$

$$=(z_1^{(k)}-\xi_1)\left[1-\frac{1+\sum_{i\neq 1}\frac{\sigma_i^{(k)}}{\xi_1-z_i^{(k)}}}{1+\sum_{i\neq 1}\frac{\sigma_i^{(k)}}{z_1^{(k)}-\sigma_1^{(k)}-z_i^{(k)}}}\right] \qquad \text{(by (18))}$$

$$= -(z_1^{(k)}-\sigma_1^{(k)}-\xi_1)(z_1^{(k)}-\xi_1)\frac{\sum_{i\neq 1}\frac{\sigma_i^{(k)}}{(z_1^{(k)}-\sigma_1^{(k)}-z_i^{(k)})(\xi_1-z_i^{(k)})}}{1+\sum_{i\neq 1}\frac{\sigma_i^{(k)}}{z_1^{(k)}-\sigma_1^{(k)}-z_i^{(k)}}}$$

$$= -(z_1^{(k)}-\sigma_1^{(k)}-\xi_1)(z_1^{(k)}-\xi_1)\sum_{i\neq 1}(z_i^{(k)}-\xi_i)\gamma_{i1}^{(k)}$$

where $\lim_{k\to\infty}\gamma_{i1}^{(k)} = 1/(\xi_1-\xi_i)^2$.

As results from the preceding section on the DURAND-KERNER-method

$$|z_1^{(k)}-\sigma_1^{(k)}-\xi_1| = O\left(\left(\sum_{i=0}^{n-1}|z_i^{(k)}-\xi_i|\right)^2\right)$$

so that the fourth order of (20) follows at once.

REMARK. (20) requires mainly the same arithmetic operations as the third order method (19): $n^2/2$ multiplications, $\frac{3}{2}n^2$ divisions; note, however, that two steps of the DURAND-KERNER-method (which need n^2 multiplications and n^2 divisions) could also be considered as one step of a fourth order method so that (20) (and much less still method (19)!) actually does *not* improve the computational efficiency of our version of the DURAND-KERNER-method.

If the Newton corrections $p(z_l^{(k)})/p'(z_l^{(k)})$ $(l=0(1)n-1)$ in (19^*) are computed in advance then one may improve the order of convergence of (19^*), too:

(21)
$$d_l^{(k)} = p(z_l^{(k)})/p'(z_l^{(k)}) \qquad l=0(1)n-1$$

$$z_l^{(k+1)} = z_l^{(k)} - d_l^{(k)}/\left(1-d_l^{(k)} \sum_{\substack{j=0\\ j\neq l}}^{n-1} \frac{1}{z_l^{(k)}-z_j^{(k)}+d_j^{(k)}}\right) \qquad l=0(1)n-1$$

$k=0,1,2,3,\ldots$.

If $p \in \Pi_n$ has n distinct roots then (21) is a fourth order method, too. Even more rapid convergence can be achieved by the following single step version of (21):

(22)
$$d_l^{(k)} = p(z_l^{(k)})/p'(z_l^{(k)}) \qquad l=0(1)n-1$$

$$z_l^{(k+1)} = z_l^{(k)} -$$

$$d_l^{(k)}/\left(1-d_l^{(k)} \sum_{j=0}^{l-1} \frac{1}{z_l^{(k)}-z_j^{(k+1)}} + \sum_{j=l+1}^{n-1} \frac{1}{z_l^{(k)}-z_j^{(k)}+d_j^{(k)}}\right)$$

$$l=0(1)n-1$$

$k=0,1,2,3,\ldots$.

An important *advantage* of (22) is that equality of some components $z_l^{(k)}$ $(l=0(1)n-1)$ usually does not lead to a breakdown of the iteration as is the case for (12), (19), (19^*) and (20).

NUMERICAL EXAMPLE.

Let $p(z)=z^3-2z^2-z+2 = (z-2)(z-1)(z+1)$, $z_0^{(o)}=2.2$, $z_1^{(o)}=0.9$, $z_2^{(o)}=-0.9$.

One step of the methods for the simultaneous determination of polynomial roots which are discussed here yields the following result:

	Method	(12)	(19)	(20)	(21)	(22)
Error	Order of Convergence	2	3	4	4	>4
$z_0^{(1)}-2$		$9.43_{10}-3$	$2.14_{10}-3$	$1.22_{10}-4$	$1.39_{10}-4$	$1.39_{10}-4$
$z_1^{(1)}-1$		$-1.07_{10}-2$	$-1.66_{10}-3$	$-1.89_{10}-4$	$2.71_{10}-4$	$2.52_{10}-5$
$z_2^{(1)}+1$		$1.25_{10}-3$	$6.99_{10}-5$	$7.80_{10}-7$	$3.36_{10}-5$	$2.35_{10}-7$

(This example is treated in HENRICI [15], p.540, with a single step version of (19^*), which uses circular arithmetic.)

4. *Error estimates for polynomial roots*

Various authors developed techniques for a posteriori error estimates for a given set of approximate roots. These devices either use Gershgorin's theorem (cf. SMITH [24], ELSNER [12]) or a fixed point argument resp. Rouché's theorem (cf.BÖRSCH-SUPAN [4], [5] , SCHMIDT [22], SCHMIDT-DRESSEL [23], GUTKNECHT [14]); another approach uses interval analysis (ALEFELD-HERZBERGER [2]) or circular arithmetic (GARGANTINI-HENRICI [13], HENRICI [15]).
There are numerous formulas for a priori error estimates treated in the textbooks of HENRICI [15] and MARDEN [20].
Here we indicate an approach which is discussed in detail in [26].

If $z_0,\dots,z_{n-1}$ are approximations for the roots of p then the basic step in (12), (19), (20) was algorithm (6) for the efficient computation of the quantities $\sigma_0,\dots\sigma_{n-1}$. As was pointed out in the first section it is necessary to represent p in the form

$$(23)\qquad p(z) = \sum_{i=0}^{n} a_i \prod_{j=0}^{i-1} (z-z_j)$$

by the general Horner algorithm (3). As may be easily verified (23) is equivalent to

$$p(z) = \det(zI-F_p)$$

if $a_n = 1$, where the complex nXn-matrix F_p is defined by

$$\begin{pmatrix} z_0 & 0 & 0 & \cdots & 0 & -a_0 \\ 1 & z_1 & 0 & \cdots & 0 & -a_1 \\ 0 & 1 & z_2 & 0 \cdots & 0 & -a_2 \\ \vdots & & & & & \vdots \\ 0 & & \cdots & 0 & 1 \; z_{n-1} & -a_{n-1} \end{pmatrix}$$

("generalized Frobenius companion matrix"). A root of p is an eigenvalue of F_p and vice versa. Thus an application of the Gershgorin disc theorem yields error estimates for $\xi_0,\ldots,\xi_{n-1}$; the following result (cf. [26]) generalizes a well known estimate due to CAUCHY (see MARDEN [20], p.122):

PROPOSITION 10. *If* p *is represented in the form* (23) *and if* ε *is the unique positive solution of the equation*

$$\varepsilon^n = \sum_{i=0}^{n-1} |a_i|\varepsilon^i , \tag{24}$$

then the roots of p *are contained in the union of the discs*

$$D_j := \{ z \in \mathbb{C} \mid |z-z_j| \leq \varepsilon \} \quad , \; j=0(1)n-1.$$

Any connected component of $\bigcup_{j=0}^{n-1} D_j$ *which consists of* k *discs contains exactly* k *roots of* p.

EXAMPLE. Let

$$\begin{aligned} p(z) &= (z-2)(z-1)(z+1) \\ &= 0.551 - 0.19(z+0.9) + 0.2(z+0.9)(z-0.9) \\ &\quad + (z+0.9)(z-0.9)(z-2.2) \end{aligned}$$

Proposition 10 then yields the rough estimates

$$\begin{aligned} & |z + 0.9| < 0.975 \\ \text{or} \quad & |z - 0.9| < 0.975 \\ \text{or} \quad & |z - 2.2| < 0.975 \end{aligned}$$

for any zero z of p. These bounds may be improved e.g. by techniques similar to those used by SCHMIDT [22]; one should observe, however, that the computational work necessary to get sharp estimates usually exceeds the one that is necessary to apply one of the above methods for the simultaneous improvement of the approximate roots to a considerable amount. We therefore adopt the point of view that it is reasonable to invest the major part of the computational work into the improvement of the approximate zeros and to rest content with rough estimates which are provided e.g. by proposition 1o (see also the remark following (14)).

Actually (24) should not be solved exactly in practice: any reasonable upper bound for ε will do it as well.

5. *References*

[1] O. ABERTH: Iteration methods for finding all zeros of a polynomial simultaneously. Math.Comp. 27, 339-344 (1973)

[2] G.ALEFELD, J.HERZBERGER: Einführung in die Intervallrechnung B.I.Wissenschaftsverlag, Mannheim, 1974

[3] G.ALEFELD, J.HERZBERGER: On the convergence speed of some algorithms for the simultaneous approximation of polynomial roots, SIAM J.Numer.Anal. 11, 237-243 (1974)

[4] W.BÖRSCH-SUPAN: A posteriori error bounds for the zeros of polynomials, Numer.Math. 5, 38o-398 (1963)

[5] W.BÖRSCH-SUPAN: Residuenabschätzung für Polynom-Nullstellen mittels Lagrange-Interpolation, Numer.Math. 14, 287-296 (1970)

[6] D.BRAESS, K.P.HADELER: Simultaneous inclusion of the zeros of a polynomial, Numer.Math. 21, 161-165 (1973)

[7] D.BRAESS, H.SPÄTH: Maßnahmen zur globalen Konvergenzerzwingung beim Newton'schen Verfahren für spezielle nichtlineare Gleichungssysteme, ZAMM 47, 4o9-41o (1967)

[8] P.BYRNEV, K.DOCHEV: Certain modifications of Newton's method for the approximate solution of algebraic equations, Zh.Vych.Mat. 4, 915-92o (1964)

[9] K.DOCHEV, L.ILLIEF: Über Newton'sche Iterationen, Wiss.Z.TH Dresden 12, 117-118 (1963)

[1o] E.DURAND: Solution numérique des équations algébraique (tome 1), Masson, Paris, 196o

[11] L.W.EHRLICH: A modified Newton method for polynomials, Comm.ACM 1o, 1o7-1o8 (1967)

[12] L.ELSNER: A remark on simultaneous inclusions of the zeros of a polynomial by Gershgorin's theorem, Numer.Math. 21, 425-427 (1973)

[13] I.GARGANTINI, P.HENRICI: Circular arithmetic and the determination of polynomial zeros, Numer.Math. 18, 3o5-32o (1972)

[14] M.GUTKNECHT: A posteriori error bounds for the zeros of a polynomial, Numer.Math. 2o, 139-148 (1972)

[15] P.HENRICI: Applied and computational complex analysis, John Wiley, New York, 1974

[16] A.S.HOUSEHOLDER: The numerical treatment of a single nonlinear equation, McGraw-Hill, New York, 197o

[17] A.S.HOUSEHOLDER: The theory of matrices in numerical analysis, Blaisdell, New York, 1964

[18] I.O.KERNER: Ein Gesamtschrittverfahren zur Berechnung von Nullstellen von Polynomen, Numer.Math. 8, 290-294 (1966)

[19] I.O.KERNER: Simultaneous displacement of polynomial roots if real and simple, Comm.ACM, 273, (1966)

[20] M.MARDEN: Geometry of polynomials, AMS Math.Surveys 3, 1966

[21] L.M.MILNE-THOMPSON: The calculus of finite differences, Macmillan, London, 1933

[22] J.W.SCHMIDT: Eine Anwendung des Brouwer'schen Fixpunktsatzes zur Gewinnung von Fehlerschranken für Näherungen von Polynom-nullstellen, Beiträge zur Numerischen Mathematik 6, 158-163, (1977)

[23] J.W.SCHMIDT, H.DRESSEL: Fehlerabschätzungen bei Polynom-gleichungen mit dem Fixpunktsatz von Brouwer, Numer.Math. 10, 42-50 (1967)

[24] B.T.SMITH: Error bounds for the zeros of a polynomial based upon Gershgorin's theorems, J.Assoc.Comput.Mach. 17, 661-674 (1970)

[25] K.WEIERSTRASS: Neuer Beweis des Satzes, dass jede ganze rationale Funktion einer Veränderlichen dargestellt werden kann als Product aus linearen Funktionen derselben Veränder-lichen, Ges.Werke 3, 251-269

[26] W.WERNER: A generalized Frobenius normal form and some applications, submitted for publication

Vol. 787: Potential Theory, Copenhagen 1979. Proceedings, 1979. Edited by C. Berg, G. Forst and B. Fuglede. VIII, 319 pages. 1980.

Vol. 788: Topology Symposium, Siegen 1979. Proceedings, 1979. Edited by U. Koschorke and W. D. Neumann. VIII, 495 pages. 1980.

Vol. 789: J. E. Humphreys, Arithmetic Groups. VII, 158 pages. 1980.

Vol. 790: W. Dicks, Groups, Trees and Projective Modules. IX, 127 pages. 1980.

Vol. 791: K. W. Bauer and S. Ruscheweyh, Differential Operators for Partial Differential Equations and Function Theoretic Applications. V, 258 pages. 1980.

Vol. 792: Geometry and Differential Geometry. Proceedings, 1979. Edited by R. Artzy and I. Vaisman. VI, 443 pages. 1980.

Vol. 793: J. Renault, A Groupoid Approach to C*-Algebras. III, 160 pages. 1980.

Vol. 794: Measure Theory, Oberwolfach 1979. Proceedings 1979. Edited by D. Kölzow. XV, 573 pages. 1980.

Vol. 795: Séminaire d'Algèbre Paul Dubreil et Marie-Paule Malliavin. Proceedings 1979. Edited by M. P. Malliavin. V, 433 pages. 1980.

Vol. 796: C. Constantinescu, Duality in Measure Theory. IV, 197 pages. 1980.

Vol. 797: S. Mäki, The Determination of Units in Real Cyclic Sextic Fields. III, 198 pages. 1980.

Vol. 798: Analytic Functions, Kozubnik 1979. Proceedings. Edited by J. Ławrynowicz. X, 476 pages. 1980.

Vol. 799: Functional Differential Equations and Bifurcation. Proceedings 1979. Edited by A. F. Izé. XXII, 409 pages. 1980.

Vol. 800: M.-F. Vignéras, Arithmétique des Algèbres de Quaternions. VII, 169 pages. 1980.

Vol. 801: K. Floret, Weakly Compact Sets. VII, 123 pages. 1980.

Vol. 802: J. Bair, R. Fourneau, Etude Géometrique des Espaces Vectoriels II. VII, 283 pages. 1980.

Vol. 803: F.-Y. Maeda, Dirichlet Integrals on Harmonic Spaces. X, 180 pages. 1980.

Vol. 804: M. Matsuda, First Order Algebraic Differential Equations. VII, 111 pages. 1980.

Vol. 805: O. Kowalski, Generalized Symmetric Spaces. XII, 187 pages. 1980.

Vol. 806: Burnside Groups. Proceedings, 1977. Edited by J. L. Mennicke. V, 274 pages. 1980.

Vol. 807: Fonctions de Plusieurs Variables Complexes IV. Proceedings, 1979. Edited by F. Norguet. IX, 198 pages. 1980.

Vol. 808: G. Maury et J. Raynaud, Ordres Maximaux au Sens de K. Asano. VIII, 192 pages. 1980.

Vol. 809: I. Gumowski and Ch. Mira, Recurences and Discrete Dynamic Systems. VI, 272 pages. 1980.

Vol. 810: Geometrical Approaches to Differential Equations. Proceedings 1979. Edited by R. Martini. VII, 339 pages. 1980.

Vol. 811: D. Normann, Recursion on the Countable Functionals. VIII, 191 pages. 1980.

Vol. 812: Y. Namikawa, Toroidal Compactification of Siegel Spaces. VIII, 162 pages. 1980.

Vol. 813: A. Campillo, Algebroid Curves in Positive Characteristic. V, 168 pages. 1980.

Vol. 814: Séminaire de Théorie du Potentiel, Paris, No. 5. Proceedings. Edited by F. Hirsch et G. Mokobodzki. IV, 239 pages. 1980.

Vol. 815: P. J. Slodowy, Simple Singularities and Simple Algebraic Groups. XI, 175 pages. 1980.

Vol. 816: L. Stoica, Local Operators and Markov Processes. VIII, 104 pages. 1980.

Vol. 817: L. Gerritzen, M. van der Put, Schottky Groups and Mumford Curves. VIII, 317 pages. 1980.

Vol. 818: S. Montgomery, Fixed Rings of Finite Automorphism Groups of Associative Rings. VII, 126 pages. 1980.

Vol. 819: Global Theory of Dynamical Systems. Proceedings, 1979. Edited by Z. Nitecki and C. Robinson. IX, 499 pages. 1980.

Vol. 820: W. Abikoff, The Real Analytic Theory of Teichmüller Space. VII, 144 pages. 1980.

Vol. 821: Statistique non Paramétrique Asymptotique. Proceedings, 1979. Edited by J.-P. Raoult. VII, 175 pages. 1980.

Vol. 822: Séminaire Pierre Lelong-Henri Skoda, (Analyse) Années 1978/79. Proceedings. Edited by P. Lelong et H. Skoda. VIII, 356 pages, 1980.

Vol. 823: J. Král, Integral Operators in Potential Theory. III, 171 pages. 1980.

Vol. 824: D. Frank Hsu, Cyclic Neofields and Combinatorial Designs. VI, 230 pages. 1980.

Vol. 825: Ring Theory, Antwerp 1980. Proceedings. Edited by F. van Oystaeyen. VII, 209 pages. 1980.

Vol. 826: Ph. G. Ciarlet et P. Rabier, Les Equations de von Kármán. VI, 181 pages. 1980.

Vol. 827: Ordinary and Partial Differential Equations. Proceedings, 1978. Edited by W. N. Everitt. XVI, 271 pages. 1980.

Vol. 828: Probability Theory on Vector Spaces II. Proceedings, 1979. Edited by A. Weron. XIII, 324 pages. 1980.

Vol. 829: Combinatorial Mathematics VII. Proceedings, 1979. Edited by R. W. Robinson et al.. X, 256 pages. 1980.

Vol. 830: J. A. Green, Polynomial Representations of GL_n. VI, 118 pages. 1980.

Vol. 831: Representation Theory I. Proceedings, 1979. Edited by V. Dlab and P. Gabriel. XIV, 373 pages. 1980.

Vol. 832: Representation Theory II. Proceedings, 1979. Edited by V. Dlab and P. Gabriel. XIV, 673 pages. 1980.

Vol. 833: Th. Jeulin, Semi-Martingales et Grossissement d'une Filtration. IX, 142 Seiten. 1980.

Vol. 834: Model Theory of Algebra and Arithmetic. Proceedings, 1979. Edited by L. Pacholski, J. Wierzejewski, and A. J. Wilkie. VI, 410 pages. 1980.

Vol. 835: H. Zieschang, E. Vogt and H.-D. Coldewey, Surfaces and Planar Discontinuous Groups. X, 334 pages. 1980.

Vol. 836: Differential Geometrical Methods in Mathematical Physics. Proceedings, 1979. Edited by P. L. García, A. Pérez-Rendón, and J. M. Souriau. XII, 538 pages. 1980.

Vol. 837: J. Meixner, F. W. Schäfke and G. Wolf, Mathieu Functions and Spheroidal Functions and their Mathematical Foundations Further Studies. VII, 126 pages. 1980.

Vol. 838: Global Differential Geometry and Global Analysis. Proceedings 1979. Edited by D. Ferus et al. XI, 299 pages. 1981.

Vol. 839: Cabal Seminar 77 – 79. Proceedings. Edited by A. S. Kechris, D. A. Martin and Y. N. Moschovakis. V, 274 pages. 1981.

Vol. 840: D. Henry, Geometric Theory of Semilinear Parabolic Equations. IV, 348 pages. 1981.

Vol. 841: A. Haraux, Nonlinear Evolution Equations- Global Behaviour of Solutions. XII, 313 pages. 1981.

Vol. 842: Séminaire Bourbaki vol. 1979/80. Exposés 543–560. IV, 317 pages. 1981.

Vol. 843: Functional Analysis, Holomorphy, and Approximation Theory. Proceedings. Edited by S. Machado. VI, 636 pages. 1981.

Vol. 844: Groupe de Brauer. Proceedings. Edited by M. Kervaire and M. Ojanguren. VII, 274 pages. 1981.

Vol. 900: P. Deligne, J. S. Milne, A. Ogus, and K.-Y. Shih, Hodge Cycles, Motives, and Shimura Varieties. V, 414 pages. 1982.

Vol. 901: Séminaire Bourbaki vol. 1980/81 Exposés 561-578. III, 299 pages. 1981.

Vol. 902: F. Dumortier, P.R. Rodrigues, and R. Roussarie, Germs of Diffeomorphisms in the Plane. IV, 197 pages. 1981.

Vol. 903: Representations of Algebras. Proceedings, 1980. Edited by M. Auslander and E. Lluis. XV, 371 pages. 1981.

Vol. 904: K. Donner, Extension of Positive Operators and Korovkin Theorems. XII, 182 pages. 1982.

Vol. 905: Differential Geometric Methods in Mathematical Physics. Proceedings, 1980. Edited by H.-D. Doebner, S.J. Andersson, and H.R. Petry. VI, 309 pages. 1982.

Vol. 906: Séminaire de Théorie du Potentiel, Paris, No. 6. Proceedings. Edité par F. Hirsch et G. Mokobodzki. IV, 328 pages. 1982.

Vol. 907: P. Schenzel, Dualisierende Komplexe in der lokalen Algebra und Buchsbaum-Ringe. VII, 161 Seiten. 1982.

Vol. 908: Harmonic Analysis. Proceedings, 1981. Edited by F. Ricci and G. Weiss. V, 325 pages. 1982.

Vol. 909: Numerical Analysis. Proceedings, 1981. Edited by J.P. Hennart. VII, 247 pages. 1982.

Vol. 910: S.S. Abhyankar, Weighted Expansions for Canonical Desingularization. VII, 236 pages. 1982.

Vol. 911: O.G. Jørsboe, L. Mejlbro, The Carleson-Hunt Theorem on Fourier Series. IV, 123 pages. 1982.

Vol. 912: Numerical Analysis. Proceedings, 1981. Edited by G. A. Watson. XIII, 245 pages. 1982.

Vol. 913: O. Tammi, Extremum Problems for Bounded Univalent Functions II. VI, 168 pages. 1982.

Vol. 914: M. L. Warshauer, The Witt Group of Degree k Maps and Asymmetric Inner Product Spaces. IV, 269 pages. 1982.

Vol. 915: Categorical Aspects of Topology and Analysis. Proceedings, 1981. Edited by B. Banaschewski. XI, 385 pages. 1982.

Vol. 916: K.-U. Grusa, Zweidimensionale, interpolierende Lg-Splines und ihre Anwendungen. VIII, 238 Seiten. 1982.

Vol. 917: Brauer Groups in Ring Theory and Algebraic Geometry. Proceedings, 1981. Edited by F. van Oystaeyen and A. Verschoren. VIII, 300 pages. 1982.

Vol. 918: Z. Semadeni, Schauder Bases in Banach Spaces of Continuous Functions. V, 136 pages. 1982.

Vol. 919: Séminaire Pierre Lelong – Henri Skoda (Analyse) Années 1980/81 et Colloque de Wimereux, Mai 1981. Proceedings. Edité par P. Lelong et H. Skoda. VII, 383 pages. 1982.

Vol. 920: Séminaire de Probabilités XVI, 1980/81. Proceedings. Edité par J. Azéma et M. Yor. V, 622 pages. 1982.

Vol. 921: Séminaire de Probabilités XVI, 1980/81. Supplément: Géométrie Différentielle Stochastique. Proceedings. Edité par J. Azéma et M. Yor. III, 285 pages. 1982.

Vol. 922: B. Dacorogna, Weak Continuity and Weak Lower Semicontinuity of Non-Linear Functionals. V, 120 pages. 1982.

Vol. 923: Functional Analysis in Markov Processes. Proceedings, 1981. Edited by M. Fukushima. V, 307 pages. 1982.

Vol. 924: Séminaire d'Algèbre Paul Dubreil et Marie-Paule Malliavin. Proceedings, 1981. Edité par M.-P. Malliavin. V, 461 pages. 1982.

Vol. 925: The Riemann Problem, Complete Integrability and Arithmetic Applications. Proceedings, 1979-1980. Edited by D. Chudnovsky and G. Chudnovsky. VI, 373 pages. 1982.

Vol. 926: Geometric Techniques in Gauge Theories. Proceedings, 1981. Edited by R. Martini and E.M.de Jager. IX, 219 pages. 1982.

Vol. 927: Y. Z. Flicker, The Trace Formula and Base Change for GL (3). XII, 204 pages. 1982.

Vol. 928: Probability Measures on Groups. Proceedings 1981. Edited by H. Heyer. X, 477 pages. 1982.

Vol. 929: Ecole d'Eté de Probabilités de Saint-Flour X – 1980. Proceedings, 1980. Edited by P.L. Hennequin. X, 313 pages. 1982.

Vol. 930: P. Berthelot, L. Breen, et W. Messing, Théorie de Dieudonné Cristalline II. XI, 261 pages. 1982.

Vol. 931: D.M. Arnold, Finite Rank Torsion Free Abelian Groups and Rings. VII, 191 pages. 1982.

Vol. 932: Analytic Theory of Continued Fractions. Proceedings, 1981. Edited by W.B. Jones, W.J. Thron, and H. Waadeland. VI, 240 pages. 1982.

Vol. 933: Lie Algebras and Related Topics. Proceedings, 1981. Edited by D. Winter. VI, 236 pages. 1982.

Vol. 934: M. Sakai, Quadrature Domains. IV, 133 pages. 1982.

Vol. 935: R. Sot, Simple Morphisms in Algebraic Geometry. IV, 146 pages. 1982.

Vol. 936: S.M. Khaleelulla, Counterexamples in Topological Vector Spaces. XXI, 179 pages. 1982.

Vol. 937: E. Combet, Intégrales Exponentielles. VIII, 114 pages. 1982.

Vol. 938: Number Theory. Proceedings, 1981. Edited by K. Alladi. IX, 177 pages. 1982.

Vol. 939: Martingale Theory in Harmonic Analysis and Banach Spaces. Proceedings, 1981. Edited by J.-A. Chao and W.A. Woyczyński. VIII, 225 pages. 1982.

Vol. 940: S. Shelah, Proper Forcing. XXIX, 496 pages. 1982.

Vol. 941: A. Legrand, Homotopie des Espaces de Sections. VII, 132 pages. 1982.

Vol. 942: Theory and Applications of Singular Perturbations. Proceedings, 1981. Edited by W. Eckhaus and E.M. de Jager. V, 363 pages. 1982.

Vol. 943: V. Ancona, G. Tomassini, Modifications Analytiques. IV, 120 pages. 1982.

Vol. 944: Representations of Algebras. Workshop Proceedings, 1980. Edited by M. Auslander and E. Lluis. V, 258 pages. 1982.

Vol. 945: Measure Theory. Oberwolfach 1981, Proceedings. Edited by D. Kölzow and D. Maharam-Stone. XV, 431 pages. 1982.

Vol. 946: N. Spaltenstein, Classes Unipotentes et Sous-groupes de Borel. IX, 259 pages. 1982.

Vol. 947: Algebraic Threefolds. Proceedings, 1981. Edited by A. Conte. VII, 315 pages. 1982.

Vol. 948: Functional Analysis. Proceedings, 1981. Edited by D. Butković, H. Kraljević, and S. Kurepa. X, 239 pages. 1982.

Vol. 949: Harmonic Maps. Proceedings, 1980. Edited by R.J. Knill, M. Kalka and H.C.J. Sealey. V, 158 pages. 1982.

Vol. 950: Complex Analysis. Proceedings, 1980. Edited by J. Eells. IV, 428 pages. 1982.

Vol. 951: Advances in Non-Commutative Ring Theory. Proceedings, 1981. Edited by P.J. Fleury. V, 142 pages. 1982.

Vol. 952: Combinatorial Mathematics IX. Proceedings, 1981. Edited by E. Billington, S. Oates-Williams, and A.P. Street. XI, 443 pages. 1982.

Vol. 953: Iterative Solution of Nonlinear Systems of Equations. Proceedings, 1982. Edited by R. Ansorge, Th. Meis, and W. Törnig. VII, 202 pages. 1982.